Successful Building Using Ecode

Sustainable Cities Research Series

ISSN: 2472-2502 (Print)
ISSN: 2472-2510 (Online)

Book Series Editor:

Bruno Peuportier

Centre for Energy Efficiency of Systems, MINES Paristech, Paris, France

Volume 2

Successful Building Using Ecodesign

Christophe Gobin
Scientific Advisor ESTP/IRC

CRC Press
Taylor & Francis Group
Boca Raton   London   New York   Leiden

CRC Press is an imprint of the
Taylor & Francis Group, an **informa** business

A BALKEMA BOOK

CRC Press/Balkema is an imprint of the Taylor & Francis Group, an informa business

First issued in paperback 2021

© 2019 Taylor & Francis Group, London, UK

Originally published in French
'Réussir une construction en éco-conception' by Christophe Gobin
© 2010 Presses de Mines, Paris

Typeset by MPS Limited, Chennai, India

Library of Congress Cataloging-in-Publication Data

Names: Gobin, Christophe (Builder), author.
Title: Successful building using Ecodesign / Christophe Gobin.
Description: First edition. | London : CRC Press/Balkema, an imprint of the Taylor & Francis Group, an Informa Business, [2018] | Series: Sustainable cities research series, ISSN 2472-2502 ; volume 2 | Includes bibliographical references.
Identifiers: LCCN 2018031743 (print) | LCCN 2018033067 (ebook) | ISBN 9780429449017 (ebook) | ISBN 9781138543232 (hardcover : alk. paper)
Subjects: LCSH: Sustainable construction. | Product life cycle.
Classification: LCC TH880 (ebook) | LCC TH880 .G57 2018 (print) | DDC 720/.47–dc23
LC record available at https://lccn.loc.gov/2018031743

Published by: CRC Press/Balkema
                Schipholweg 107C, 2316 XC Leiden, The Netherlands
                e-mail: Pub.NL@taylorandfrancis.com
                www.crcpress.com – www.taylorandfrancis.com

ISBN-13: 978-1-138-54323-2 (hbk)
ISBN-13: 978-1-03-209465-6 (pbk)
ISBN-13: 978-0-429-44901-7 (ebk)

Table of contents

Introduction

Ecodesign:
Why?
What for?
In what conditions?

When the concept of ECODESIGN first emerged in the building industry, some feared that it would only add confusion to the subject of sustainable development.

This methodology, which consists in "considering environmental aspects right from the design stage" is not so recent, and was introduced into the industrial sector over fifteen years ago. However, such an observation does little to reassure numerous professionals who consider that the building sector is too specific to take advantage of industrial methods.

The advantages of this proactive approach are, however, easier to understand when placed in the perspective of now obvious environmental challenges. A more detailed analysis of these challenges helps us to understand all the benefits of taking a new approach, and in particular to evaluate potential in the light of practices used in other contexts.

Ecodesign nevertheless involves more than a simple set of tools. It responds to a number of basic principles that need to be applied in the specific case of production and our living environment. Assessments of this kind are not neutral, either, since they underline certain practices that have often lost their meaning through routine production methods.

This critical exercise is the only way of setting out in detail how the ecodesign process is applied in the different building phases. Although this work requires a degree of rigour, it provides a fairly rich approach that can be applied to the different projects that make up our built environment.

Climate challenge: the risk of global warming.

The 21st century began with a major debate on the environmental risks at large, and in particular climate change.

Increasing amounts of greenhouse gases resulting from industrial activity are accumulating, reducing the earth's protection from the sun's rays. This modification is set to gradually increase temperatures with irreversible consequences if measures are not taken soon.

The issue is the object of scientific discussion, yet manifestations of the phenomenon are multiplying in the form of perturbed sea currents, cyclones, dwindling ice caps, gradually shrinking mountain glaciers, and increasing desertification.

Beyond drawing up a base list, a more important, non-controversial task is to define the scale, or more precisely the perimeter concerned. All of these observations concern the planet, they are not restricted to specific geographic zones. They are taking place all over the world and raise global questions, even though the causes are not always exactly the same.

It is even probable that all of these manifestations are only a more tangible translation of the systemic way that our contemporary world works, which has been overlooked for several decades. We all hold a stake in a very fragile ecosystem that is highly sensitive to the exogenous variations caused by our "artificially" created activities.

> *Buildings constitute a microclimate that is itself affected by global climate change.*

This climate risk on a global scale also specifically concerns individual building projects in at least two different ways through changes in the weather.

Global warming, with the higher incidence of hot periods in our temperate regions, brings up the question of summer comfort. Some general contractors envisage going beyond established practices, and call for technical solutions that can cope with higher average summer temperatures and so reduce periods of discomfort (i.e. days on which a given temperature is exceeded).

The presence of wind brings up similar questions. The number of storms resulting from local microclimate phenomena is increasing and intensifying, implying that the resistance of some building parts (especially roofs) needs to be reviewed. This phenomenon is generally coupled with higher rainfall, with the risk of inopportune flooding.

The combination of these two phenomena brings a new problem, which is the level of water tables, which affects the resistance of the ground. This issue is particularly critical for buildings with superficial foundations in which geotechnical risks have emerged. It is particularly prevalent in suburban housing zones resulting from urban sprawl.

Multiple issues therefore need be tackled, and resolving them means applying a new attitude in each individual building project.

> *The energy challenge: the risk of dependence and depletion.*

The issue that ranks second is energy. Its decreasing availability is initially experienced through a considerable drop in spending power. However, in addition to this economically perceptible effect, two deeper trends are at work.

The first of these, the best illustration of which is oil, is the depletion of resources, which is at the origin of most of the increase in energy costs. The drop in available resources has clearly provoked a steep rise in prices. It has also led to more costly prospection (i.e. very deep, at sea, etc.). Some specialists think that the market justifies seeking more expensive supply sources, like bituminous sands. However, these means of production raise concerns about environmental impact and their treatment necessarily brings additional costs.

The second trend has a geopolitical dimension. Reserves are not only less and less promising, they are also concentrated in specific regions generally located far from Europe. The issue of dependence therefore arises. Although users tend to ignore it, there is a genuine risk of an ill-timed shortage of energy supplies. Global governance could be set up in the long term, but not within the next few years, while reserves of other fuels (gas, uranium, coal) are also limited. In any event, it will only be effective if there is a genuine reduction in the volumes used.

All analyses point to an inevitable quota system to meet the energy needs of our collective and individual activities.

Buildings are a source of considerable energy consumption, both for users and the community.

When it comes to construction, it is worth bearing in mind that the sector represents 40% of natural resource withdrawals on a global scale. As an example, a 63 m^2 apartment made of concrete mobilizes around sixty tonnes of materials (aggregates, steel, cement, etc.), or one tonne per useable square metre.

This usage raises questions for strictly local reasons. In several French regions, the extraction of some aggregates is now subject to quotas or even banned. More generally, taxes on extraction have recently been raised. Some European countries go as far as imposing high taxes on the use of non-recycled materials.

However, this only concerns resources mobilized to create a built environment. Yet the question is relevant throughout the operation phase, during which significant volumes of fluid (water, fuel oil) and energy (electricity, gas) are consumed. The built environment also makes a global contribution of 40% of greenhouse gas emissions (in France, due to the nuclear energy policy, this figure is only 20%).

Given these conditions, it appears difficult to build without considering this aspect of "withdrawing" resources. In fact, construction should be envisaged with a restricted view of its procurement capacities. Some experts are starting to consider the notion of finiteness as an inevitable constraint that should condition all decisions, including in construction work.

Financial challenge: the risk of insolvency.

The environmental issues set out above directly concern the ecosystem as an ecological system. However, an additional question concerns the means of remediation, or in other words the capacities to create a regulation that could put a stop to the causes of imbalance.

Any talk of resource depletion implies in the long term the emergence of what some experts call the "environmental debt". Historically, this corresponds to the withdrawal of natural resources by colonial powers throughout the world. This should logically result in a price increase. The globalization phenomenon neutralizes this process to some extent, yet it appears unavoidable in the long term. The result will be even greater efforts to get access to these raw materials.

However, a second surplus cost arises from treating externalities. Until now, numerous incidences of pollution have not been taken into account and their treatment has been postponed or delayed. In a perspective of respecting the environment, this work will need to be financed. The means for doing so have not yet been defined, but the so-called "polluter pays" principle appears the most logical. With the aim of striking a financial balance, this would necessarily involve covering corresponding expenditure or an increase in the price of products.

The conjunction of these two phenomena results in a dilemma: whether to spend more and thus diminish funding capacities, or innovate intensely to obtain better yields from our techniques. The risk of collective insolvency leads some movements of opinion to say that we should move towards obligatory degrowth. For others, on the contrary, it provides a genuine opportunity for extending technological efforts and making progress.

> *Viewing construction as a resource rather than a handicap leads us to try and optimize its holding cost.*

The building industry cannot ignore this "controversy". We might even reasonably ask ourselves whether it is not already underway.

Numerous professionals consider that increasing buildings' energy performance necessarily increases the technical cost. Many reason that individual objects would need to have their quality upgraded. In such a case they would not be wrong. However, an alternative does exist, which is to respect new performance by redesigning the object. This "design to cost" practice is familiar to some industrial sectors that are subject to very strong outside competition.

On a macro-economic level, the cost of construction is also a subject of concern. Clearly, the supply of buildings is insufficient, but their price is also too high to open up wider access. In these conditions, a building risks becoming a handicap instead of fulfilling its primary function, which is to act as a support for all the activities of society.

In this sense, construction is at the heart of the different environmental issues envisaged from a societal point of view.

> *Ecodesign seeks to manage the flows generated by buildings.*

Faced with the issues mentioned above, ecodesign is defined as an approach that seeks to minimize the different environmental impacts of each object.

In the case of buildings, this means thinking about the causes that generate impacts and reducing them as much as possible. This entails four principles that guide the use of ecodesign.

The first of these consists in considering a construction as a "thermodynamic machine", i.e. a system that mobilizes resources and produces waste. Preserving the environment naturally means reducing the volume of withdrawals and discharges/ pollution waste, etc.

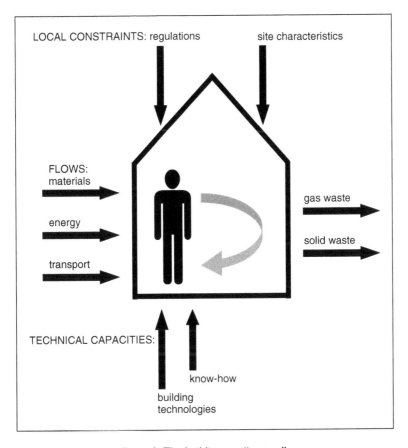

Figure 1 The building as a "system".

A building could therefore be represented by the above diagram, taking into account local constraints and technological capacities.

An ecodesign building is capable of producing environmental accounting illustrating the impact of using resources.

Ecodesign is therefore based on the idea that buildings operate as systems, transforming inflows into outflows. As a result, an appraisal of the result must take into account users' expectations of usage conditions.

It is quite normal that the inside comfort temperature required by an occupant leads to higher energy flows.

Thus, ecodesign presupposes that the construction on which it is based already matches the purposes of the end user. Put another way, it necessarily applies to a "functional unit" whose perimeter must be fully defined.

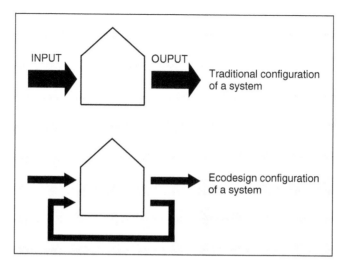

Figure 2 Visualization of the result of ecodesign: The objective of reducing flows should not result in modifying usage performances. In addition, impacts are not simply correlated to the volume of flows, but also depend on the type of materials used.

This observation is crucial, since it differs from the traditional status of buildings, which are generally considered to be static, i.e. fixed objects. Ecodesign has a dynamic vision of buildings, because it assimilates them to working mechanisms.

Ecodesign envisages all the phases of an operation's life span.

To consider a building's operating mode involves examining this behaviour over the long term. Ecodesign covers every phase of a building's life cycle.

This requirement introduces two parameters into the analysis process: (1) the time horizon of usage, (2) a stage-by-stage breakdown of the life of the project considered.

The horizon corresponds to the building's life expectancy. This does not mean the construction's durability (i.e. the time until it becomes a "ruin"), but the duration assigned to its usage (i.e. expected usage time). This is therefore information that can be assimilated into an economic decision: for how long will this project be used from the end user's point of view?

The breakdown into phases corresponds to the building's availability. It thus starts with its construction and ends with its demolition, which some describe as the journey from "womb to tomb" (i.e. from start to finish). Each of the four phases observed includes intermediate stages set out in detail in the diagram below.

Ecodesign therefore brings together two timelines: one transcribes the different usage periods, the other sets out the necessary tasks for carrying out these different usages.

Ecodesign compares building solutions using an "LCA" (Life Cycle Analysis) that visualizes the value of twelve impacts. Since it involves

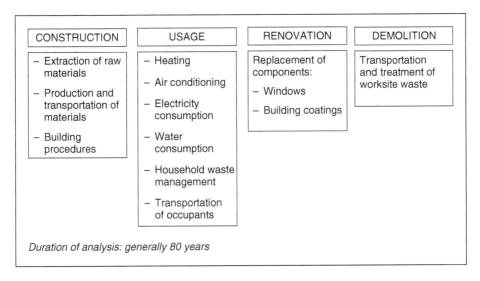

CONSTRUCTION	USAGE	RENOVATION	DEMOLITION
– Extraction of raw materials – Production and transportation of materials – Building procedures	– Heating – Air conditioning – Electricity consumption – Water consumption – Household waste management – Transportation of occupants	Replacement of components: – Windows – Building coatings	Transportation and treatment of worksite waste

Duration of analysis: generally 80 years

Figure 3 Different stages making up the phases of a building's life.

a multi-criteria analysis, each solution corresponds to a compromise and is never the only solution in a given context.

Ecodesign measures the quantities corresponding to the different flows generated during each phase of the lifespan.

It then transforms these quantities into impact values using conversion matrices that define each entry's contributions to the different issues, of which there are currently twelve or fourteen in the "neighbourhood" LCA.

These impact factors correspond to scientific expertise established by European, or international, research networks. They are therefore protocols conveying a certain state of the art that is more or less consensual depending on the items examined.

This calculation is even more interesting when it is used to compare two building scenarios. A comparison that involves superposing two radar charts provides information that enriches choice capacities depending on the importance attached to each impact.

These environmental profiles are therefore highly useful because they constitute a decision-making tool, all else being equal, because they are drawn up in the same way.

This entire procedure is called a life cycle analysis, and is now an international standard.

Ecodesign is a voluntary, responsible approach.

Ecodesign, which seeks to manage environmental impacts, can be assimilated to a risk-minimizing approach. As such, it requires responsibility from all those that implement it.

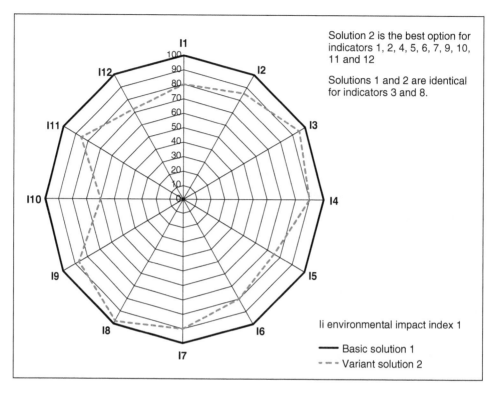

Solution 2 is the best option for indicators 1, 2, 4, 5, 6, 7, 9, 10, 11 and 12

Solutions 1 and 2 are identical for indicators 3 and 8.

Ii environmental impact index 1

——— Basic solution 1
– – – Variant solution 2

Figure 4 Environmental profile of two solutions for the same project.

However, viewed this way, a new problem arises that calls for several precautionary principles.

Until now, the traditional approach to building has made do with sequential work, with the aim of putting together a set of physical items step-by-step. This has been translated by a chain of tasks, starting with the design of the object, its justification vis-à-vis current regulations, and ending with on-site production in line with these decisions.

Ecodesign does not work in terms of an object, but rather in terms of how it operates. It considers a building as a system that interacts with end users and its environment. It cannot therefore be viewed as a simple succession of individual contributions, but requires genuine collaboration between workers in a participative co-development process. This involves progressively putting together the whole building for a single purpose rather than based on individual approaches.

> *Ecodesigning a building involves a genuine prediction study that needs to be carried out collectively and cannot be subdivided.*

Apart from rare exceptions, a construction is not a "shelf product". It makes an object available that must in the future operate in conformity with expected results.

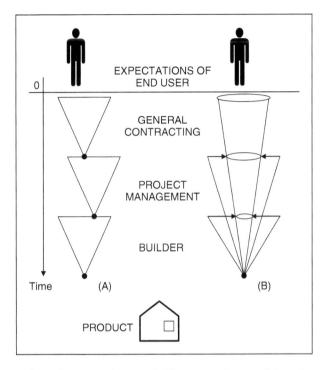

Figure 5 The traditional, sequential approach (A) compared to a collaborative approach (B).

This configuration implies that professionals responsible for delivering the operation anticipate its future usage by simulation, and then gauge appropriate technical solutions.

To do this, they need access to usage scenarios that stipulate the modalities of usage (e.g. occupation time, regulation, people present) and presumed behaviour (activities carried out, apparatus used, etc.). None of these hypotheses will be useful unless they relate actual occupation modes. At present, calculation methods are based on conventional scenarios that are very different from real-life experience. This explains the gaps observed between statutory theoretical calculations and the consumptions actually invoiced.

As a result, ecodesign requires a broad discussion of the operation modalities. This necessarily entails genuine consultation between all parties involved focusing on shared hypotheses.

Ecodesign introduces a new relationship with end users.

With the increased cost of energy, attention is focused on effective consumption, in particular for buildings. It now seems clear that actual expenditure is much higher than theoretical calculations imply.

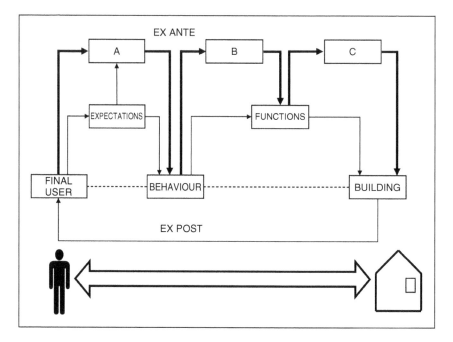

Figure 6 Buildings are artefacts that work in close relationship with users. Ecodesign fits into a construction cycle deduced from this user-living environment interaction in three stages: A- Programming, B- Design, C- Production.

Professionals are gradually discovering that the sizing of constructions is very often determined on the basis of usage hypotheses that do not reflect precise usage behaviour. Ecodesign illustrates this discrepancy and raises the question of user responsibility.

The objective is not to devise so-called "robust" solutions that would compensate the bad use of a building, but rather to debate on a new relationship with end users.

In particular, it appears indispensable to stop separating the provision phase (i.e. providing the physical object) from the operation phase (i.e. the object in operation). This implies admitting a man/machine interaction that has been recognized by other economic sectors for a long time.

> *In ecodesign, performance is guaranteed through transparent mutual commitments between stakeholders.*

Ecodesign, which considers this aspect linked to the building's operation, has been devised to differentiate several levels of the concept of performance. They are, in order of appearance:

- Nominal performance, characterizing users' perception, i.e. the result they expect in terms of usage.

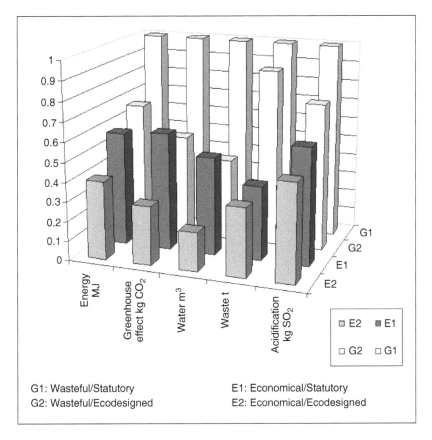

Figure 7 Incidence of end user behaviour on performance.

- Operation performance, corresponding to the criteria required for ensuring the preceding level's stability over time. This is therefore the expression of the conditions for obtaining the nominal value.
- Modalities of ownership, relating to a recently introduced class and covering all prior action carried out to make users aware and show them how to use the building. The aim of this category is to support usage and it has become necessary due to the sophistication of equipment and initial assistance for handover in conformity with the scenarios chosen to use the systems available.

These three classes can be organized using the table below, which structures expectations.

Two-tier recommendations.

Ecodesign is not a process that is superimposed on traditional practices. In fact, it adds to routine methods by introducing new decision criteria.

Table 1 Eliciting commitment to carry out an ecodesign operation in its totality.

	Synchronic approach (instant)	Diachronic approach (duration)
SPECIFICATION Commitment to result	– Nominal usage performances	– Usage scenario – Maintenance scenario
FOLLOW THROUGH Complementary action	– Awareness-raising – Display	– Operating instructions – Servicing instructions – Maintenance instructions

Ecodesign therefore involves combining a new approach to the conditions for carrying out each phase of a project with additional tasks introduced as a result of a wider field of analysis.

An ecodesign approach is only worthwhile when applied to an operation that begins by responding to the expectations of the end users. If a building is to be favourable to the environment, it must first provide the conditions required for the future activities it is set to receive.

For this reason, recommendations to develop ecodesign all start off with the reminder that projects must match their essential purpose. The aim of this preliminary analysis is not to rewrite the modalities confronting professionals, but rather to get "back to basics", i.e. the actual reason behind their work.

It is only on this basis that ecodesign can bring genuine added value when carrying out an operation.

This contribution can be broken down into several components, i.e.:

– Terminology that introduces several specific concepts.
– Reasoning adapted to pursuing environmental issues.
– Protocols for measuring the effects of the decisions made.
– Databases conveying the acquisition of shareable information.

Ecodesign never produces solutions that can be transposed from one project to another. The idea is not to obtain ready formulas but to open up a discipline and a new way of acting in which choices are gauged to fit each particular context.

Recommendations can therefore be compared to the invariants of a methodology. They are general rules that will need to be broken down for each local configuration but that will each time deal with the product aspect and process aspect of the project.

Recommendations relative to each phase of the operation's lifespan.

Ecodesign covering all of the phases in a built environment's lifespan entails appropriate treatment for each of them. The four standard phases of a building operation can be organized into two groups: phases that define the built object, followed by those concerning its provision and its usage.

Phase 1, known as programming, consists in drawing up the project specification. This requires a number of tasks calling for practitioners and specific skills.

During phase two, known as designing, these intentions are translated into a spatial form that will require different types of engineering to physically produce it on site.

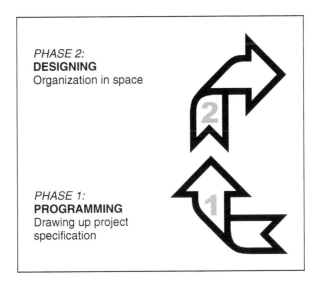

Figure 8 The first two phases correspond to preparation work ("virtual object").

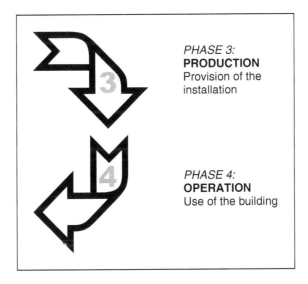

Figure 9 The other two phases concern the physical building ("actual object").

These two stages in the life of a project can be described as intellectual work that essentially involves putting together a technical package (paper in computer format).

To cover its whole life cycle, the operation needs to include two complementary stages, i.e. its production and its operation.

Phase 3, known as production, corresponds to the worksite. The planned object is constructed so as to become a working installation. This phase also results from a convergence of numerous concurrent tasks.

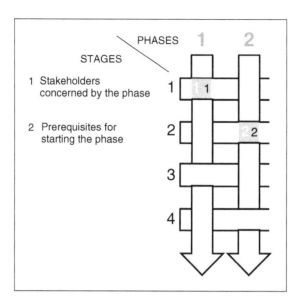

STAGES

PHASES 1 2

1 Stakeholders concerned by the phase

2 Prerequisites for starting the phase

Figure 10 The initial stages of the ecodesign process constitute the first class of recommendations.

Phase 4, known as operation, corresponds to using this object during a period that is generally fairly long. It can take very variable forms since in addition to its initial use, the object may be updated or rehabilitated, setting off a new programming phase.

These two phases last for very different amounts of time, the first from one to two years, and the second taking decades, since buildings are products with long cycles.

A preliminary scheme used to situate each recommendation.

As for all processes, each phase of the ecodesign (1 to 4) signifies the transformation of input items to provide a suitable outcome (outputs). This movement can be divided into four steps organized into two classes.

The first concerns the early stages of each phase, and can also be split into two stages: (1) mobilizing specific practitioners whose allotted task will be to accomplish this phase of the life cycle; (2) gathering essential information.

In the case of ecodesign, this preliminary work is particularly rich. These specific contributions constitute a set of recommendations. They can be viewed as the necessary conditions for producing a more environmentally friendly building.

The importance attached to these initial stages of each phase result from a well-known attitude among professionals, who associate the quality of the outcome of a task with the care taken in preparing it.

With regards to their implementation, the preceding recommendations lead up to a product (3) that must then be transmitted to the stakeholders of the following phase (4). These two new stages define the second class of recommendations.

The product, or in other words, the result obtained following the treatment specific to the phase, is characterized by renewed contributions that thus constitute the added

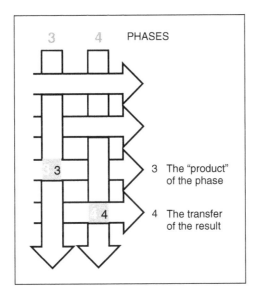

Figure 11 The finishing stages of the ecodesign process constitute the second class of recommendations.

value specific to ecodesign. It seems obvious that this methodology actually contributes to re-evaluating customary practices.

However, the benefits are highly dependent on whether professionals in the following phase are able to take on the results without feeling the need to question them because they lack confidence in the earlier work. This transmission plays a crucial role in the efficiency of the whole value chain.

Through this four-stage analysis of the different phases of a building operation, the ecodesign aspects can be tackled with a degree of logical rigour.

Chapter 1

Ecodesign in the programming phase

The general contractor is the order giver who ensures the project's financing.

1.1 MOBILIZING THOSE INVOLVED

The major figure during this phase is by nature the general contractor.

It is the general contractor who initiates and ensures the project's financing. He does this directly or indirectly depending on whether he provides the amount needed to carry out the operation out of his own funds, or whether he arranges a financial set-up for other partners whom he represents.

It is this financial role that is considered first. But it is important to remember that out of principle the general contractor also represents the future end users of the project.

Two arrangements are possible. The general contractor may want to build for himself as a future occupant or as an owner-landlord. He might also act as an inter-mediary who, once the operation is delivered, sells the building, in which case he is a promoter.

However, his essential jurisdiction is in all cases to obtain a positive property balance sheet. This involves at least a balance between the sum of the cost of the land and the future technical cost of the construction and the usable revenue. The general contractor is therefore the first to assume the right to build for a terrain.

As a result, his main goal is to get the most he possibly can out of a site's potential capacity. This optimization encourages him to make the financial dimension the primary choice criterion. This financial aspect, which obviously depends on the market economy model, is however accentuated under the effect of land price inflation.

1.1.1 The general contractor's role

But he is above all the representative of future users of the planned building. In this perspective, he is obliged to anticipate the building's later usage, which will have an incidence on its life cycle.

In ecodesign, the general contractor fulfils the same criteria as in standard practice. However, he finds himself with new responsibilities. He is no longer the sole protagonist in the construction phase, gradually calling on professionals to work for him.

In this case, a series of stakeholders have a say in the project, with no money-based contractual bond. The general contractor is obliged to take into account the greater influence of local residents, ecological associations, and pressure groups involved in community missions.

For all of these entities, the project brings different degrees of disruption. For local residents, the future building may create an inconvenience that restricts their view or alters the way the neighbourhood operates. For some, the worksite will generate diverse nuisances, although these will be temporary. Lastly, some may consider that the project upsets the site's environmental balance.

The general contractor will therefore need to find a compromise taking in all of these points of view, and his choices will be influenced by considerations that are no longer purely financial.

On a more general level, for an ecodesign, a general contractor's responsibility to society will also put the emphasis on environmental criteria. In addition to strictly respecting regulations, he may voluntarily commit to making a more significant contribution to tackling climate change and the depletion of natural resources. This commitment, which corresponds to taking a stand on environmental issues, is essential for the project's continuation, and the general contractor is the only one in a position to assume this role and impose it on all those involved in the project.

> *The end user is all too often physically and unconsciously missing in the value chain.*

In the programming phase, the second actors are the end users, although they cannot be physically present in the project since they are only definitively designated when the project is ultimately delivered.

However, traditionally in the sector, this "person" is rarely invited to participate in developments with professionals. Because end users are reputed to be incapable of expressing genuine demands, they find themselves conveniently excluded from discussions.

Put another way, although buildings are clearly destined for users, these users are bound to leave the work solely in the hands of professionals, and take possession only when the products have been built, without any tangible opportunity to make significant modifications.

In fact, end users are reduced to the socio-professional category to which they belong, with the corresponding behaviour codified via "default" regulations.

1.1.2 The end users' role

> *They are, however, the core stakeholders of all building projects, and it is on them that the ecodesign process should focus.*

In ecodesign, the relationship with end users is not fundamentally different, since their physical presence is also delayed until the project's delivery.

However, this methodology is based on a concept that puts the focus on end users. It reasons in terms of a "functional unit" that is the foundation of all measures used throughout the project.

As this expression implies, what is evaluated from an environmental aspect is the way the building operates under certain conditions of performance usages, which are directly correlated to how end users appreciate the building.

In fact, in ecodesign, the entire project is managed so as to respect expectations in terms of usage, ranging from comfortable temperature and acoustic protection to the level of lighting. All of these results, which are conditioned by technical choices, are devised to satisfy users, not in a standard way, but in order to match their wishes as closely as possible.

In these conditions, it is in the general contractor's interest to fully apprehend his request at an early stage and define performances that correspond to realistic, rather than conventional, practices. It is clear that this kind of attitude involves taking risks, but it is more responsible when it comes to the contractual relationship.

> *Drawing up a programme calls for specific, professionally established expertise.*

The third class of actors capable of intervening in this phase is that of the programme managers.

This is a recent speciality that does not yet correspond to a totally stable body, which explains why programme managers can also be architects acting as advisors, or engineers more specialized in spatial allocation.

The role of these new practitioners is to draw up the project's technical specifications that will serve as a framework for all the other phases of the operation.

This mission is more common for tertiary operations, because it involves defining needs in terms of surface area and the level of the corresponding equipment in line with specific demands.

Programme managers therefore carry out the general contractor's role by delegation. Their production consists in supplying synoptic tables that indicate for each premises its dimensions, finishing work and any services. This often very detailed description should allow architects and companies to propose constructive solutions that correspond to validated requirements.

The programme is thus an important part of the contractual relationship between the general contractor and the future project management.

1.1.3 The programme manager's role

> *This involves advance specification of the building's future usage, which will be the most important aspect in the long term, including from an environmental point of view.*

In ecodesign, the programme manager writes up a document that is more specifically directed at end users, in as much as the specifications are drawn up with users in mind.

This document will not just specify surface areas, it will stipulate the conditions for use in terms of lighting levels and thermal, acoustic and olfactory comfort (level of ventilation, etc.).

All of this information on modalities of use constitutes targets to aim for without presupposing the means that will be implemented in the future.

In fact, the programme will be a very detailed breakdown of the functional unit organized for the benefit of the end users.

Some programme managers even esteem that the final document cannot be written without long discussions with end user representatives. They view this co-authorship as a guarantee that future beneficiaries will feel totally involved in the project.

This kind of process can clearly only be envisaged for dedicated projects.

A building is by definition destined to be put to a particular use.

The general contractor starts by selecting the purpose of the building that he is planning in two ways that have gradually lost their significance, despite being crucial.

The general contractor started off by reflecting on the nature of his project, or in other words the purpose of the operation. Will it be residential, tertiary, industrial facilities, etc.? This question is not neutral when asked simply in relation to the project's context. As a general rule, it is mainly the suitability of the site that will condition this type of operation. In many ways the operation is pre-determined by local regulatory constraints (transport plans, local planning, etc.). Nevertheless, project developers who get involved in such conditions, even if primarily as a commercial venture, personally assume the purpose of their building from that point on.

Yet beyond this tacit commitment, when general contractors accept the location of their projects, they are imposing definitive, binding constraints on all of the practitioners involved. These constraints are not just physical, they are also contextual. Physically, they include the geometric (or cadastral) shape, along with sunlight and orientation conditions. Contextually, they correspond to the site's characteristics in terms of landscape, traffic, access, complementary services and proximity. All of these data enter into the choice of architectural scheme in as much as they are compatible or at least reconcilable with the choice of purpose.

The general contractor begins by working one-dimensionally, and in terms of economic theory this mode can be qualified as a macroeconomic choice.

1.1.4   Purpose of the operation

The comprehensive use of a building results from carefully reflect-ing on the built environment to make it a suitable, efficient structure, both operationally and environmentally.

In ecodesign, the purpose is also respected, although the approach is very different.

Rather than the overall purpose, the accent is on the operation underlying it. In the same programme, the types of user will significantly influence the way activities are organized. Instead of a general description, each usage will need to be taken into account.

It is indispensible to move beyond stereotypes and set out detailed expectations. Without this, the project becomes more of a "speculative operation", i.e. standardized and thus taken "as is" by the end users, who must accept it without any other possibility for modification.

The objective in ecodesign is to characterize the purpose more precisely, i.e. as a functional unit. This involves specifying usages by fixing a performance level related to the conditions offered by the building that make it easier to carry out the various activities that the building accommodates.

This means that the building is no longer considered as a static object, but as a "logistical support" available to the end users. In these conditions, what counts is the detailed operating mode that will then be evaluated.

The purpose needs to be broken down and formulated in terms of the expected functions for each activity that the building will accommodate.

> *A building's purpose is frequently assimilated to the type of future users.*

In addition to the purpose, the general contractor defines the nature of the building's users. It is worth analyzing how he does this.

The standard approach works in terms of clientele, breaking down a demand by means of purchase capacity. When the overall target is financial equilibrium, the emphasis is on the solvency of the market and segment considered.

In this case, the major criterion is to conform to a stereotype model that corresponds to standard habits considered as acceptable to the majority.

The only sector that avoids this simplification is that of equipment, which often corresponds to a particular use and thus requires specific programming. The residential and tertiary sectors, on the other hand, are subject to commonplace approaches that group expectations into a handful of classes ascertained from a marketing approach that is fairly standardized beyond useful surface area requirements.

Building classifications are therefore generally based on purpose, and broken down in terms of standards chosen by the general contractor.

1.1.5 Describing the users

> *Designating users mainly involves thinking about their behaviour vis-à-vis the building they will be taking possession of, which will influence the building's operations.*

In ecodesign, the occupants must be described more precisely, since users' behaviour conditions how well a building operates.

It is therefore preferable to define a detailed timetable for each type of user and establish how they will use the built area that they will be provided with.

This mainly takes the form of a usage scenario that stipulates the following for each significant time period:

- The number of people present,
- The comfort conditions sought (e.g. comfort temperature, settings expected, etc.)
- The equipment used.

It is only by providing these data that it will later be possible to model the responses and evaluate the result in relation to the expectations of the different stakeholders.

This information naturally requires a more detailed analysis of the future clientele of each operation. It can never be definitive because it can only be confirmed when the users actually take possession. However, fairly reliable studies are now possible based on behavioural analyses of consumers and carried out by specialized firms.

> *Programming transcribes the conjunction between the availability of a terrain and the ways in which a "new" building can fit into it.*

Traditionally, a general contractor is spurred by his own volition to undertake an operation, or by the decision of a third party that asks him to see a project through (as the contracting authority's representative).

The programming phase is most frequently the result of a prior opportunity study that has substantiated the idea of successfully constructing a building. This task mainly consists in adjusting three items, i.e.: (1) the availability of land, (2) the existence of potential customers, (3) the capacity to draw up a finance plan permitting a realistic "set-up".

Contractually, the objective is to financially succeed in making a new building available within an existing neighbourhood.

1.1.6   Pre-contractual form

> *In ecodesign, the contract takes on a new responsibility, which is to make the operation as neutral as possible in relation to the environment.*

The ecodesign approach entails an initial commitment from the general contractor that adds a new dimension to his project.

The aim is no longer simply to accomplish a building in the best financial conditions, but to contribute to creating a new living environment while minimizing the impact on the existing environment. This responsibility, which usually resides in a voluntary decision, shows how much importance the order-giver attaches to the environment, which itself comprises several scales (local, regional and global).

On a more institutional level, ecodesign comes under the decider's societal responsibility. This must thus be clearly indicated so as to bring all of the practitioners into the value chain. It will work best if it goes beyond basic regulatory constraints.

Although this kind of attitude is still relatively rare, it is gradually becoming part of a broader approach, which is to contribute to sustainable development.

1.2   CONSIDERED PARAMETERS

> *A real estate operation implies starting off with a positive financial situation.*

The feasibility of a project is initially conditioned by the parameters of its financial equilibrium. This involves grouping a certain number of data together.

The first thing to verify is the competition for the same type of building within a close distance from the future site. This review can be used to differentiate the programme and highlight its strong points when commercializing the operation.

The second question relates to the acceptable price level corresponding to potential demand in order to quickly cover the programme's financial needs.

In fact, all of the conditions required on this register relate to the market position. The nature of the programme is not as important as the customer catchment area. This observation translates the value attached to the real estate market, which was until recently very high.

The general contractor will then subtract this figure from the building land fee, to obtain the technical cost of building available for the project management. This target cost constitutes the framework for the whole value chain. However, this budget does not really have a direct link with end users.

1.2.1   Economic feasibility

But it must also balance out over its entire life cycle, which is what makes ecodesign worthwhile.

In ecodesign, the financial aspect is characterized not only by the solvency of the end user, but also by the nature of the holding cost.

As well as the initial amount required to take possession of the building, the cost of its operation, and even deconstruction, must be taken into account. This is because a building is designed to last for a certain length of time, known as its life cycle.

Ecodesign extends over the entire operation of the building and seeks to achieve an optimal operating cost. The objective is not only to make the building useable, but to make it last over time by ensuring that the costs of maintenance and operation remain within users' purchasing power.

Construction costs therefore not only include the envelope, they also take operating expenditure into consideration.

All projects fit into a context subject to constraints.

Urban-planning constraints mostly define what we might call the building's capable envelope. This expression refers to the limits that will be imposed both on the project's differentiation and geometry.

On the first point, methods are based on documents that have developed in line with awareness of changes in the urban environment. In other words, a town should be organized to take into account real-life experience and not just imposed policies (c.f. revisions of land use conditions).

Concerning volumetry, constraints result from rules on building outlines devised to foster overall consistency in shapes and construction lines, as well as on the issue of sunlight or shadows cast on buildings close by.

Another crucial element that conditions the set-up of any operation relates to the site's location and situation. The location will naturally determine whether the building is connected up to different types of networks. The situation should be considered as a factor in the land's development.

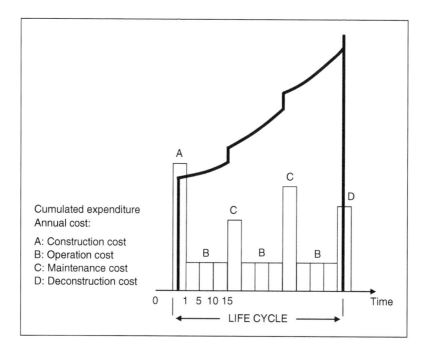

Figure 1.1 The main components of "overall cost".

The building land fee also weighs on the project balance and often explains the urban sprawl resulting from a search for cheaper building terrains.

1.2.2 Feasibility vis-à-vis urban planning

A project cannot stand alone – it necessarily interacts with its environment, on a physical or societal level. Environmental profiles take this into account.

In ecodesign, urban planning is primarily viewed as an issue of scale.

Buildings are not self sufficient – their successful operation mostly depends on the project's relationship with its environment, i.e. services close by, transport networks, cultural facilities, etc. It is therefore worth thinking carefully about inter-dependence before making decisions about the envisaged programme.

In particular, ecodesign pays close attention to the transit generated by the building's insertion into the urban fabric. Increased traffic caused by the project has an immediate incidence on the site's road system. And it can also contribute to a rise in greenhouse gases if users are obliged to make longer journeys than before.

Urban planning is thus considered as a systemic vision of the living environment.

Sustainable development requires at a minimum that a building satisfy the site's so-called environmental criteria.

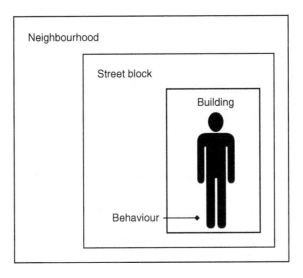

Figure 1.2 How the different scales of an operation interlock.

Taking sustainable development into account is a fairly recent approach that now seems unavoidable.

In terms of programming, the general contractor can proceed in two ways when it comes to sustainable development. The first of these is to strictly comply with the restrictions of the moment, and the second is to take a more proactive, personal approach.

Externally, municipalities increasingly make their own initial diagnosis of local sustainable development issues. This usually takes the form of an Agenda 21. This document constitutes a framework to be applied to all new real estate programmes.

In general, the minimum constraint is to impose so-called environmental certification on buildings. Given the actual structure of these labels, this usually takes the form of fairly slack obligations that boil down to setting up an operation management system (OMS).

Nevertheless, several communities go further and impose stricter thermal performance levels than the regulations. In such cases, they either grant more extensive building rights (up to 20%) or reduce the building land fee if they control it (e.g. promoter competition by reversed auction).

The general contractor draws up his programme in line with this context.

1.2.3   Feasibility vis-à-vis sustainable development

Respecting the environment also means controlling as far as possible the impacts resulting from the building's availability and operation.

Ecodesign deliberately concerns only the environmental dimension of sustainable development. However, it aims to be operational and puts the emphasis on measurements, with the idea that the general contractor will be clear about his targets and able to evaluate them without simply providing a statement of principle.

For this reason, an ecodesign approach sets only a limited number of performance indicators and promotes significant dialogue between all of the stakeholders on a project.

Five indicator classes are selected that are subject to a protocol of measures in line with French and European standards (respectively P01.20.3 and CEN TC 350).

Projects are always located on specific sites.

Buildings are primarily physical objects of significant size. Those who design them need to take into account a number of context-specific factors. These can be classed into three families corresponding to: (1) the plot as a land surface that will host the operation (2) the zone above this surface, i.e. the climate, and (3) the zone below it, i.e. the subsoil.

The following table provides an overview of the main characteristics that also serve as input data for drawing up the envisaged programme. This stage therefore involves gathering information, and the care allotted to this task will condition whether an operation runs smoothly in the future. All of this information should be prerequisite to the drafting phase.

Box 1.1: Six environmental indicators that illustrate a sustainable development approach.

Withdrawals

Volume of natural resources used
P1

Volume of energy consumed
P2

Volume of water consumed
P3

Discharges

Volume of green house gas emitted
R1

Volume of waste generated
R2

Human toxicity
R3

Table 1.1 Terminology of physical data relating to the site.

Climate	Data on annual temperatures
	Data on precipitation (pluviometry)
	Solar radiation
	Data on dominant winds
Plot	Topographical survey
	Distributors/Connections
	Road networks
	Biodiversity (fauna, flora)
	Neighbourhood
Subsoil	Geotechnical surveys
	Nature of water tables
	Underground networks

1.2.4 Collecting physical data

The aim is to obtain the best result while taking specific features into account. Specialists call this a bioclimatic approach.

In ecodesign, physical data are of considerable importance. The objective is not to consider these elements as constraints, but rather to view them as specific advantages and make good use of them.

As an example, we can take the way the plot is treated.

To optimize the location's topography, the focus is not so much on the landscape itself as on the opportunities it brings, such as its capacity to provide shelter from the wind, the possibility of benefiting from free solar gains, or taking advantage of a scenic view.

This approach can be extended to all the other elements characterizing the site on which the project is located.

Projects are subject to a series of rules established by the community.

An operation can only be planned in accordance with its complete immediate environment, from both a physical and regulatory point of view.

This latter aspect, unlike for urban planning, is more specifically centred on the detailed rules that apply to the project and with which the building will need to conform.

These might be considered as the "protocols" on which the construction will need to be justified, or in other words the modalities that direct its sizing. Over and above the institutional side of these elements, there must be a concern for the community or its representatives to ensure that the operation at a minimum respects all the conditions required to guarantee a degree of continuity for the building's living environment. These are the common building rules.

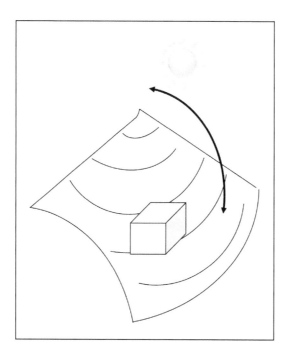

Figure 1.3 The physical elements that ecodesign focuses on: – dominant winds, – views, – solar exposure.

Table 1.2 Terminology of regulatory data relating to the site.

Context	Agenda 21
	Sustainable development plan (PADD)
	Land use plan (PLU)
Geometry	Construction lines
	Size
	Building rights
Performance	Particular specifications
	Energy performance
Codes	Thermal regulations
	Snow and wind regulations
	Eurocodes (structures)
	Fire regulations
	Sanitation code
	Acoustics

1.2.5 Analyzing regulatory data

However, rules should be thought through and are not necessarily considered as sufficient. In ecodesign, they may be intentionally reinforced.

Ecodesign goes further than simply respecting obligatory regulations. The objective is not to conform to rules, but to define satisfaction levels as closely as possible.

This observation requires a more detailed explanation that takes standard practices into account. Regulations clearly apply to all, yet they correspond to a principle of precaution. Designated representatives of community life have established rules because they cannot know whether individual general contractors are capable of making choices concerning the community.

When taking a responsible approach, general contractors must ask themselves whether such obligations are sufficient given the project's specific context, or whether it is preferable to be more demanding in view of particular issues.

This kind of attitude, which is at the core of ecodesign, is never easy to adopt because it introduces additional constraints vis-à-vis the immediate financial balance. However, over a longer period, this decision can turn out to be a useful anticipation. A good illustration is the rapid changes in thermal regulations. Should a building due for delivery in 2012 (BBC level) but commenced in 2009 (RT 2005 level) be satisfied that it fulfils regulations on the date of the building permit, even though its life expectancy is thirty years or more?

Programmes transcribe the expectations of end users.

For the general contractor, knowledge of future users is a precondition to drawing up a programme. This knowledge takes the form of three types of information that need to be provided.

The nature of users defines the building's main purpose, but it also stipulates future users' specific features. This includes gender composition, age groups, routines and even tastes. Lifestyle trends may also be considered.

The activities carried out by users give a more precise characterization of expectations in terms of earmarked areas. Rather than relying on stereotypes, all too often translated by occupation ratios, a more detailed analysis of daily timetables indicates the actual usefulness of conventional areas.

User relations can be listed so as to determine the nature of common areas outside the individual sphere. This concerns all vertical and horizontal movement and access points. It also differentiates areas used for strictly private purposes from those that include shared activities.

1.2.6   Knowledge of users

They need to be defined very precisely in order to establish significant, measurable expectations.

In ecodesign, the information-collecting process needs to go much further, not in an intrusive way, but rather as an indispensible condition to anticipating the ways in which the building will be used in the future.

The characterization of comfort is highly dependent on individual users. For example, a so-called comfortable temperature, or one perceived as comfortable by a user, cannot be reduced to the standard regulatory threshold for his or her life style. The same observation applies to lighting levels and even useful surface areas.

In addition to the above-mentioned data come conditions for adjusting atmospheres, which are indispensable for at least two reasons. One is that the ways people use a building may evolve over a long period (e.g. behaviour changes, ageing), and in particular external conditions may change (e.g. weather, climate change). It is thus worth ascertaining the (admissible) tolerance level for each condition of use.

1.3   OBTAINED RESULTS

Three types of activity are traditionally singled out and identified (in inventories).

Private activities: These correspond to the individual actions personally undertaken by each user. To be carried out satisfactorily, they require indispensible geometral conditions (e.g. useful volume to work, rest, have fun, etc.).

Semi-private activities: As their label implies, these activities involve several users, but in low numbers, based on tacit understanding (e.g. family activities in residential buildings or group work activities in tertiary buildings).

Collective activities: These are much more random; they are often fortuitous in as much as the identity of those involved is not predetermined. Such activities generally result from meetings between users of the building and "others" in the community (e.g. receiving visitors, meetings, discussions, etc.).

1.3.1   List of activities

They also need to be put into a long-term perspective.

In ecodesign, activities are analyzed from two points of view. The first is to verify the activity's durability, and the second consists in envisaging the activity in terms of its environmental impact.

Progression of activities over time: On the one hand, activities are never static; they change in line with a schedule (e.g. daily timetable). In addition, they can transform over time under the effect of different parameters (e.g. modified uses, economic conditions).

Regrouping of activities by zone: Given that financial capacities are necessarily limited, it is not possible to devote an area to each different activity. As far as possible, they will need to be grouped with a multi-purpose approach.

These two analyses make it possible to anticipate the impact that the building's operation will have on the environment.

A building corresponds to an operating choice and physically represents a certain number of assumptions.

By nature, buildings are physical objects providing a venue in which individual activities can take place in the best conditions for those undertaking them.

In fact, buildings provide a means that makes up for the lack of available surface areas in urban zones. This corresponds to the creation of an "artificial environment"

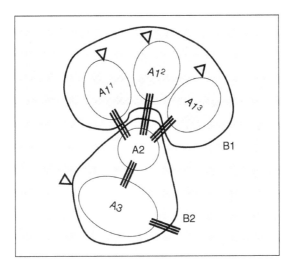

Figure 1.4 Analysis of activities by nature: A1: private activities; A2: shared activities; A3: collective activities. Analysis of activities by zone: B1: specialized area; B2: multi-purpose area.

Table 1.3 Terminology of programming principles.

Satisfy physiological pre-requisites	– Provide a space – Provide an atmosphere
Manage interactions with the environment	– Ensure protection – Use goods and tools – Manage relations
Make available conditions for accomplishment	– Fit into a site – Carry meaning

that some people qualify as a habitat by extension of the term used in natural sciences to qualify a species' biotope or ecosystem.

In these conditions, buildings are a means to satisfy several criteria set out in the table 1.3.

1.3.2 Operating principles

The main objective is to make better use of accessible resources to make the building available.

In ecodesign, the main concern is to respond to the core principles of the project while limiting the amount of resources used.

Box 1.2: Terminology of ecodesign principles.

Reduce requirements (reduce consumption)	– Reduce surface areas and volumes used – Reduce withdrawals at source (grey energy, quantities of material, water savings) – Promote restraint (behaviour) – Increase yields – Optimize occupation duration – Foster responsible use – Reduce additional transport flows
Use renewable resources	– Take advantage of free gains – Favour renewable energy sources (e.g. wind, sun), rainwater, agro-based materials
Encourage recycling	– Re-use materials – Recover losses – Recycle waste
Preserve the ecosystem	– Contribute to improving public health

Table 1.4 Terminology of nominal performances linked to the use of a building.

Space	– Geometric dimensions – Size – Additional load required
Atmosphere	– Comfort temperature – Level of lighting – Interior acoustic level – Volume ventilated per hour
Protection	– Safety of people (fire, accident, etc.) – Security (vandalism) – Sun and health protection
Goods & tools	– Nature of equipment – Capacity of use – Acoustic level
Relations	– Level of access control (independence) – Level of protection vis-à-vis outside noise
Site	– Nature of services available – Level of environmental regulations
Semiology	– Nature of image given to centres – Capacity to make it one's own

The way a building operates is first defined by performance criteria that characterize a level of response to the expectations of end users.

At all times, end users expect buildings to provide them with a location suitable for their activities. To achieve this, it is indispensible to characterize a building by its usage, in other words for the following performances.

1.3.3 Level of nominal performance

These performances, which must be stable over time, can only be achieved under certain environmental conditions.

In ecodesign, nominal performances need to be viewed in the perspective of long-term operations. Buildings are thus considered as thermodynamic machines whose yield must be optimized.

Buildings operate in a similar way to machines. The conditions of use required constitute their operating performances.

Buildings have fairly long lifespans. Nominal performances, which must be maintained at a constant level during this period, necessarily involve maintaining and upgrading buildings throughout their life cycle.

Box 1.3: Terminology of so-called possession performances (usage over time).

Withdrawal level	To achieve performance: – volume of materials – grey energy mobilized To maintain performance: – annual consumption – resources for maintenance
Waste level	Contribution to greenhouse gases Production of waste: – solid waste – liquid waste
Adaptation level	Capacity to improve nominal performances Capacity for flexibility, adjustment to variations in activity Capacity to change purpose
Deconstruction level	Capacity to re-use materials Capacity to recover resources

The main reason is the unavoidable physical ageing of all of a construction's components. This is the result of two phenomena:

1 Natural wear and tear
2 Consequences of technical obsolescence.

These two evolutions more or less coincide depending on the nature of the building parts, which have different cycles.

In addition to the performances experienced by the end user, it is therefore important to consider the conditions in which this result is obtained since new resources will be mobilized.

In his programme, the general contractor includes detailed information on the thresholds to be respected and the frequency of interventions. These operating performances are added to the consumption required to work the various pieces of equipment that contribute to usage performances.

1.3.4 Operating performances

Thinking about a building's operation means reflecting on the way it is used over time because it interacts with users.

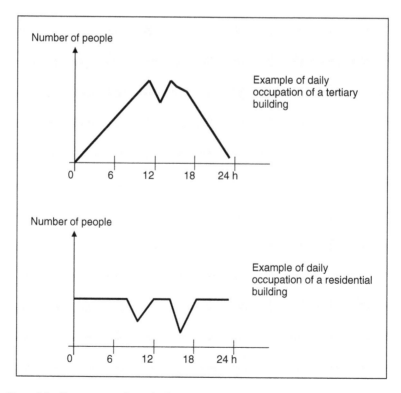

Figure 1.5 Comparison of two building occupation scenarios for different activities.

In ecodesign, operation is taken into account systemically and in a fairly detailed manner.

In addition to the above-mentioned concerns, it is important to understand how a building's occupants will behave. One reason is the effect that behaviour has on buildings' energy requirements. This therefore involves defining lifestyles and using them in simulations, rather than making do with a standard scenario drawn up as a legal precaution rather than to transcribe a realistic mode of occupation that can be used by the general contractor.

Whether these data are reliable remains questionable. However, when it comes to future energy performance contracts, a reliable modelling of behaviour is indispensible in the long term and will come under the responsibility of the general contractor.

> *Given the accelerated rhythm of change in society, it is vital that general contractors stipulate the probable lifespan of their buildings.*

One of a general contractor's prerogatives is to look closely at the lifespan of the building planned. Buildings are frequently intended to operate for several decades. The general contractor should therefore announce the usage period of his project.

This attitude is still fairly rare, since it overrides a "default" procedure implicitly contained in calculation codes. In France, the justification rules for rough work in a building presume a very long life span. However, some contemporary buildings have much shorter life cycles.

As a result, it is essential to establish the probable lifespan in order to draw up a reliable economic balance sheet that can be used to make informed choices independently from the character of the buildings' durability. Here, the term durability means in the absence of ruin.

The distinction between these two terms should be made in particular when the status of the installations changes. The durability was attached to the asset value. When this asset value starts to make way for a usage value, then the difference lies in the lifespan, which is linked to the capacity to offer the same level of performance over a certain number of years.

For the general contractor, the notion of lifespan is therefore an appreciation of the risk of obsolescence. This analysis is particularly useful in the case of tertiary projects confronted with rapid changes in organizing work teams or office equipment in the building.

1.3.5 The life cycle horizon

> *A life cycle necessarily covers a precise policy on support for buildings.*

In ecodesign, lifespans are analyzed at several levels. A building programme thus specifies a system that is considered in its totality.

Although a building's overall operation is characterized by its lifespan, this lifespan is made up of components with different natures, each with their own specific lifespan. The equipment in a building ages faster than the structure. Finishing elements are refreshed much more frequently. Technical choices can be optimized in line with the declared lifespan.

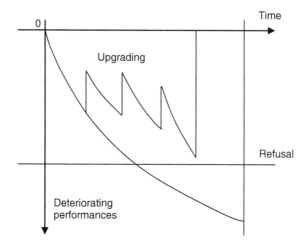

Figure 1.6 How the life cycles of the different parts of a building link together.

Along with lifespan, the general contractor thus needs to stipulate the preferred future policy on maintenance and upkeep.

This information does not constitute a definitive commitment, but it should make it possible to draw up a response in terms of a plausible scenario that for all practical purposes characterizes the order-giver's involvement in his operation.

> *The budget allocated to an operation establishes a framework for other practitioners, but its explicit announcement shows that the general contractor is taking responsibility for the project.*

A programme that defines all the characteristics of an operation should also determine its budget.

This depends on two elements:

1 The outcome of the feasibility study,
2 The level of specified performances.

It would be possible to supply only the specifications and rely on project management to verify whether the final building fits in with the projected budget. However, this would mean simply repeating established practices.

A well-reasoned target cost reveals an incentive to look for innovative solutions. Particular emphasis should be laid on the fact that specifications designate the expected results without ever stating (setting out) the means.

The general contractor's role, in fact, is not to obtain the lowest cost, but to optimize the performances of his project within the limits of the resources that he is able to mobilize.

The announcement of a programme's budget therefore indicates a genuine command of the function of order-giver.

1.3.6 Objective cost

In these conditions, ecodesign is a constrained optimization approach.

In ecodesign, the key criteria is not economic, since the main aim is to master the consequences that making the project available will have on the environment.

However, an ecodesign approach contributes to project savings, at least in terms of the energy used for operating the building.

The main question is to determine whether higher thermal performance results in higher initial building costs than continuing with less ambitious technological routines.

This could be the case if approach criteria were applied independently from a matter of principle. However, this approach needs to be closely linked to all of the programme choices, and perhaps even preside over them. In such conditions, it is an extremely powerful lever for innovation.

1.4 TRANSFER MODALITIES

The graphic form of a programme is an operation chart.

All of the principles that the general contractor selects for his operation can be represented graphically in the form of an operation chart.

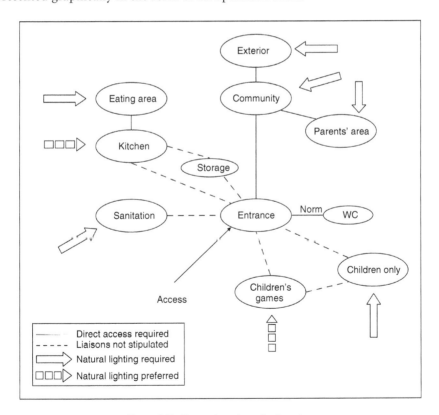

Figure 1.7 Operation chart for housing.

This figure is a coded representation showing how the various zones end users employ for their activities fit together and their relationships in terms of proximity and frequency of use.

It gives an indication of operations that need to be organized in space, and as such does not prejudice a geometrical form. However, it corresponds to a well-reasoned decision that involves the whole operation.

The representative codes are not standardized but are established on a case-by-case basis. The aim is to indicate the general part of the project.

1.4.1 Operation chart

The indispensible reference for all future evaluations.

Ecodesigns are based on this document. It is the first material expression of what is known as the "functional unit" and what will be optimized from an environmental impact point of view.

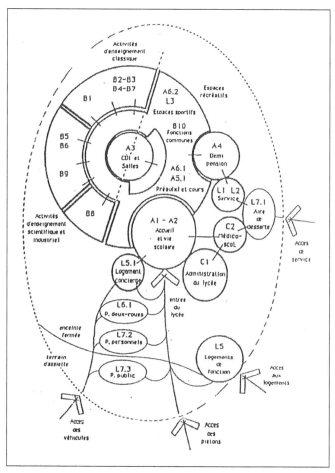

Figure 1.8 Operation chart of a college.

By analogy with computers, an operation chart constitutes an "object model" that sets out the different components of the operation considered as a system and how they are connected.

All of which shows the importance of this conceptual representation at the origin of the building process. The foreword of norm P01-20-3 (evaluation of buildings' environmental performance) highlights the necessity of this reference document.

Programmes constitute a road map for all future practitioners.

The graphic documents drawn up by the general contractor need to be supplemented with a consolidation at the level of the performances linked to each activity zone. This work is delivered in a literal form, i.e. word list tables.

It is worth a reminder that this task should be based on specifications rather than recommendations, since it concerns an expression of expectations and not resources.

The quality of this stage resides in the comprehensiveness of the data shown. These must stipulate all of the contracting choices, although some of them will be left open to decisions made by future practitioners.

The structure of the data is fairly rich, as illustrated by the following list:

Box 1.4: Contents of an operational programme.

PROGRAMME:
1. *Activities chosen*
1.1 Private activities:
   – List of activities
   – Indication of duration
1.2 Semi-private activities:
   – List of activities
   – Indication of duration
1.3 Group activities:
   – List of activities
   – Indication of duration

2. *Expected performances:*
2.1 Principle per activity:
   – Operational choices selected for the project
2.2 Performances:
   – List of performances
   – Indication of measurement
2.3 Operating conditions:
   – Indication of upkeep and maintenance choices

3. *Operation chart:*
3.1 Private zones
3.2 Community zones
3.3 Transition zones

1.4.2   Structure of an application for tender

This should be as precise as possible to avoid future challenges due to lack of information and support the effects of ecodesign.

From an ecodesign point of view, to continue with the computer analogy, the performances selected constitute the object attributes

Each zone of the chart is characterized by a precise list of the different performances required to provide suitable support for end users' activities.

Strictly speaking, each performance level should be linked to the tolerance range considered acceptable to ensure correct usage. However, the building industry is not yet mature enough to introduce such demanding standards. The values shown are therefore targets to be aimed at as closely as possible.

This description also extends to the budget allocated, both in terms of financial resources and time to availability.

The table below provides an illustration.

Box 1.5: Additions to operating specifications.

GENERAL DATA:

1.1   Data on the site:
1.1.1 Plot:
 – Geometric characteristics
 – Morphology and climate
1.1.2 Sub-soil:
 – Geotechnical survey
 – Geology & hydrology
1.1.3 Environment:
 – Networks/concessions
 – Neighbouring services

1.2   Constraints:
1.2.1 General regulations:
 – Land use plans, etc.
 – Zones NV, TH, earthquake, etc.
 – Noisy or sensitive sites
1.2.2 Land planning constraints:
 – Common ownership
 – Size
 – Constraints on materials

1.3   Target customers:
1.3.1 Typology:
 – Customer characteristics
 – Probable developments
1.3.2 Life styles
 – Activities and usage
 – Solvency

1.4   Project organization:
1.4.1 Practitioners:
 – Prop. Devepr, Proj. Manager, Tech. Engineer, controls
 – Contractual relations
1.4.2 Phases:
 – Project stages
 – Controls carried out
1.4.3 Time frame:
 – General envelope
 – Time per sequence

1.5   Resources allocated:
1.5.1 Investment
1.5.2 Operating expenditure

> *To be respected, a programme has to be credible, i.e. realistic given the specific parameters of its context.*

In terms of responsibility, the general contractor cannot simply set out the expected outcomes. He must also verify that the specification is compatible with financial resources available to successfully see the operation through.

Prior validation of the programme's transfer can be done in two ways:

1   by verifying that performances are consistent with each other,
2   by ensuring the existence of response capacities.

Consistency implies genuine professional experience in setting performance levels that are not contradictory. In particular, some comfort hypotheses have a direct incidence on operating costs. The same is true for acceptable thresholds of useful surface area per type of activity.

In fact, and especially to ensure financial feasibility, the best solution is to simulate an initial architectural response and verify a rapid estimation of it. This approach makes it possible to "fine tune" the programme and ensure that it is compatible with the expertise available.

This last point brings up an additional question, i.e., knowing who to contact to respond to the programme. The high technical level of building work naturally requires specialist project management, even though many practitioners, in particular architects, rail against this segmentation.

The level of performance required therefore needs to be linked to the programme's ambition vis-à-vis future professionals that can potentially be mobilized.

1.4.3 Validation criteria

> *This reliability test extends to the environmental component.*

In ecodesign, the above reasoning is also valid, but is extended to cover environmental performances. However, given that technical expertise is less developed than in other disciplinary fields, it is necessary to use other procedures.

The most common practice is to use "best practices", i.e. past references. This involves making general use of feedback on experience, which is not frequently done in France. Nevertheless, some thresholds are beginning to be understood, often through comparison with European studies.

It is not impossible, however, to carry out a prior Life Cycle Analysis (LCA) based on a very early plan in order to situate the challenges and validate selected objectives. The EQUER software programme can be used for this kind of exercise.

However, performances relating to the operating phase can be substantiated using some databases put together by groups of users, in particular in the tertiary sector.

> *All too often, programmes express an intention that requires confirmation by the project that is to be offered in response.*

The programme established by the general contractor constitutes the specifications that determine the project management mission.

However, this document, which sets out the different aspects of the planned construction, is generally a prescription. It groups factual data in a very descriptive manner, in other words it essentially states surface area requirements, without stipulating the corresponding operating mode.

The accent is on geometry and on size, implicitly taking the conditions and usage scenarios to be "conventional". This term groups both the building regulations and conventions that by default substitute a detailed observation of user behaviour.

In its traditional form, the programme is therefore a collection of data that in principle constitute the starting point for the next stage, which determines how the surface area requirements will be organized in space. This document, which materializes the end of the general contractor's active role working solo, is a milestone in the project's proceedings; however, it does not strictly speaking constitute a breakpoint since it is often subject to interpretation by the general contractor.

This is all the more so since the programme will serve as a basis for launching the architecture competition that will lead to designating the project management.

I.4.4   Post-contractual form

> *In ecodesign, the status of the programme is much firmer and is defined as a target that validates the reality of expectations.*

The originality of ecodesign is that it transmits a "performance-based" package to the future project manager. By this term we mean that the expected results of the overall operation are designated. There are two major advantages in this approach:

In defining his expectations, the general contractor not only takes total responsibility for his choices vis-à-vis future users, he also calls on the expertise of other practitioners to bring him appropriate responses at a register in which he lacks the competences. This dialogue is thus marked by genuine confidence: both from the order giver, who has an objective but calls specialists for help, and from the project manager, who will be able to put his talents to use in a stable, thought-out programme.

In addition, this dual relationship allows each of the protagonists to work within a well-defined framework of responsibility. The general contractor remains within his role, fixing the performance level to be attained on all registers that correspond to future activities carried out in the planned building. The project manager alone determines the most suitable architectural and technical solutions. This division of roles makes it easier to recognize skills and ensure that they are complementary.

However, there is a third advantage to carrying out a "performance-based" programme, i.e., the possibility of having a verifiable reference source throughout the project. Defining performance levels makes it possible to carry out measures, or in other words, to be capable of evaluating results as they develop.

A performance is only of interest if it can be measured. If this observation is applied, it introduces a virtuous circle linking all those working on a project.

It is no use a general contractor writing up a detailed programme if he cannot use it to verify whether the project handed over to him at the end of the building work is in conformity.

The advantage of the programme is that it effectively fixes measurable performances at the building's delivery. This approach, known as "commissioning", is slowly becoming inevitable, even though it is still not practised in France.

However, if it is to become general practice, the question of measurement needs to be addressed first. If performances are requested, how will they be measured?

All nominal usage performances can be measured, although doing so is often complex. Yet it is possible to verify comfort temperature, lighting levels, acoustic attenuation, etc. For mechanical performance, consensus exists on calculation simulations and the resulting justifications.

All of this information concerns what users experience in situ. In fact, the spatial configuration will have been accepted at a much earlier stage in the architectural project agreement during the EPA phase. The same goes for everything relating to regulation conformity (e.g. fire protection, human security, etc.).

Added to measurement questions comes the issue of responsibility and potential compensation in case of non-observance of the results indicated.

1.4.5 Verification conditions

Environmental impacts are more complex to quantify, but simulation tools can determine the reasons for any possible deviations.

Ecodesign makes use of two different types of measurement.

The volume of withdrawals and grey energy, i.e. the energy required to manufacture the components of a building, are easy to check based on the quantities actually implemented.

Energy consumption during the operating phase is much harder to measure. Although invoices quantify the output of operations, the causes of the volume consumed are highly diverse and random. They depend heavily on meteorological variations and they also translate the actual behaviour of users. Although external temperatures can be monitored, behaviour is difficult to piece together. What can be measured, however, is the effective performance of the building's envelope.

In fact, methods currently exist that manage to define performances in a significant way by combining tests on the airtightness of façades with monitoring of consumption over short periods during winter and summer. These methods use measuring processes that are increasingly accessible thanks to progress made in terms of meteorology (measuring flows).

All of these apparatus can be used to rapidly attain a good command of results, since each of them can be measured in real time.

1.4.6 Synopsis

A programme provides a detailed description of the individual relationships that the building makes it possible to establish between end users and external environments (i.e. other stakeholders).

These expectations define a reference that remains constant throughout the planned operation.

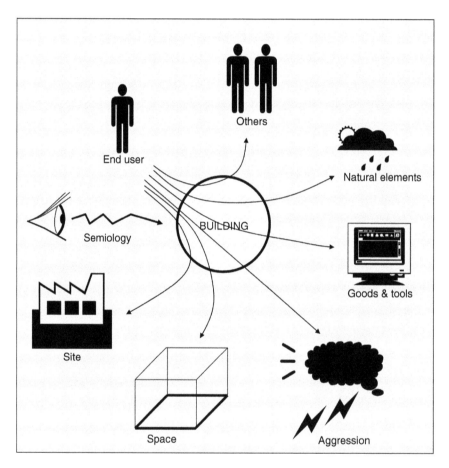

Figure 1.9 A programme is a matrix of the functional unit, which is the ecodesign reference ...

1.5 OVERVIEW DIAGRAM

The main advantage of ecodesign is that it supplements the criteria already taken into account in the functional analysis. In particular, it makes it easier to determine each operation's environmental performances.

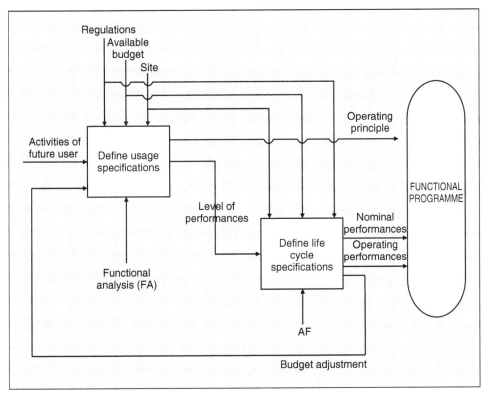

Figure 1.10 ... It is the result of a process that responds to a logic: the functional analysis that is itself a protocol between all of the stakeholders.

Chapter 2

Ecodesign in the design phase

Can the figure of the architect be anything else than a "deus ex machina"?

2.1 MOBILIZING THE ACTORS

The architect is at the heart of this phase of the operation, which involves transcribing the programme in the form of a project.

The architect's role is to give substance to an intention that becomes apparent from reading the specifications and from contact with the general contractor. The next step is creation. This act is naturally personal and partly results from experience acquired over several years at art school.

In France, this exercise must obligatorily be carried out by an entity that is free to make its own choices and does not depend on a financial structure that could be involved in the operation's proceedings. This provision fulfils a concern to make architecture a collective requirement independent from any interest. This status has led to several patterns of conduct.

The project decision therefore finds itself focused on a single person, even though it is the result of teamwork between the architect and his collaborators. This is the result two combined reasons. All too often, a programme is incomplete and left to the initiative of the project manager, who is considered as skilled in the art. In addition, the outcome of the project is assimilated to an artistic work that cannot be put into question without the approval of its author.

This creates a two-fold focus on the project design phase in which: (1) the general contractor expects an image and (2) the project manager hatches the project. As a result, this stage takes on such a level of intensity (some might say catharsis) that all the following stages follow on from each other with the sole perspective of accomplishing "The" project and much less with the concern of responding to end users. This observation, which may seem harsh, is more understandable taking the view that the optimization effort is not collective, but contracted out downstream.

2.1.1 The role of architectural project management

Architects are like conductors directing a music score written into a building's programme.

In ecodesign, the architect still plays a primary role, but he operates in a different context.

The programme is more fully developed and establishes a precise framework. The architect works under constraint. On the theoretical level of architectural design, this mode of creation is not restrictive and in no way holds back the imagination. However, it is more demanding, since the objective is to refine the content of the operation's specifications in a spatial way. The architect puts himself at the service of the project.

The composition work is organized around a single concern, which is that of resolving the equation set by attaining the expected performance level, which is viewed as the successful expression of in-depth preliminary work carried out on the use of the planned construction.

Ecodesign leads the architectural project manager to optimize all of his choices. This is what Paul Valéry made clear in his book *Introduction to the Method of Leonardo da Vinci*, when he wrote, "… that slow transformation of the notion of space which started off as a vacuum chamber, as an isotropic volume, and has gradually become a system inseparable from the matter it contains and from time."

Which illustrates the importance of respecting the functional unit in ecodesign.

> *The role of technical project management is to ensure the feasibility of the architectural project.*

Along with the file produced by the architectural project management, the pre-sizing written into the project's geometry needs to be validated.

This is because the sizes defined in the architect's plans need to be justified on a technical level. This task covers aspects of mechanical resistance, thermal and acoustic behaviour. It involves verifying that the building in its draft configuration corresponds to the different calculation codes practised by the profession as a whole.

This latter remark is not strictly speaking a standard formula. It simply highlights the fact that the common approach, especially for regular buildings, is restricted to

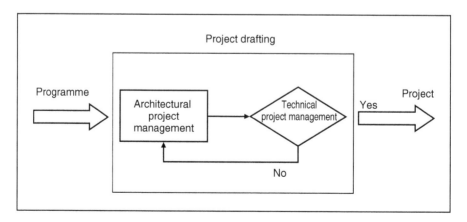

Figure 2.1   In a traditional process, the role of technical project management is to validate the draft project.

convention, i.e. the application of a protocol considered as a consensus on the state of the art at a given point in time. Thus, the rules for calculating reinforced concrete changed significantly from 1970 to 2010, moving from BA 68 to BAEL and finally to a Eurocode. Each time the objective is to improve precision.

2.1.2 The role of technical project management

The most appropriate configuration for getting the most out of ecodesign is concurrent engineering.

In ecodesign, the two project management sides do not have fundamentally different roles – the difference is how they work together.

Given that the building must satisfy much more precise specifications, the work can no longer be sequential, but concomitant.

To guarantee the assigned performance levels, much closer collaboration must be put in place starting from the rough layout. The reason for this is simple: the project is a succession of iterations that must converge towards an optimum that satisfies both the architectural intentions and a standard operating mode.

A codevelopment approach is ultimately much more natural, given the complexity of a construction considered as a system. This mutual enrichment can take very different forms, ranging from a preliminary questionnaire to group work on the "same site". The aim is to make sure that the choices made are effective vis-à-vis the use that will be made of the building in the future.

Savings in an operation often result from adjusting both expectations and responses.

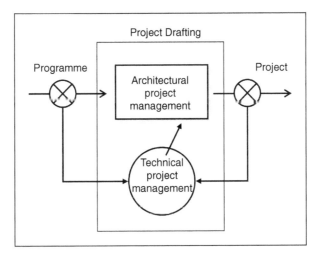

Figure 2.2 In ecodesign, the two project management teams must work together to codevelop the project.

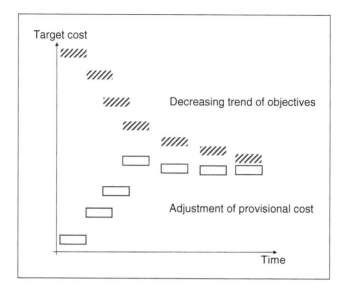

Figure 2.3 Progressive convergence of economic and functional objectives in the project designing phase by successively reducing the differences.

All projects are established in line with a budget identified by the project management. The economist's role is determined by this requirement. It guarantees that the cost of the building is achievable within the limits set.

However, it is important to remember that the notion of cost is very relative, since it depends on numerous variation factors. These are, for example, building habits, mass production effects, learning curves, market conditions, etc. In fact, a technical cost translates at a given point in time an agreement between the different stakeholders on how to resolve a technical problem.

This balancing mechanism between ambitions and possibilities can be illustrated by the diagram below. In a traditional approach, a compromise is reached following progressive developments that consist in looking for "savings" both on a technical level and regarding the project's ambitions.

It is interesting to note that this dynamic process is maintained right until the end of the call for tender on the building work.

2.1.3 The economist's role

> *The logical objective here consists in determining the most efficient solution at a constant cost.*

In ecodesign, the approach is totally different for two basic reasons.

Ecodesign does not deal with the financial aspect at all. For each situation, it determines the level of control of environmental impacts, all else being equal. In particular,

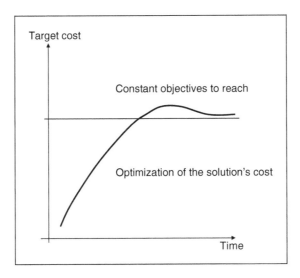

Figure 2.4 The continual process constituted by the design of a product with a designated objective.

the technical cost of the selected choice comes under the sole responsibility of the project management team.

However, ecodesign makes it possible to determine the best choice between two solutions with the same cost. It applies a dual-constraint optimization: the chosen solution must respect the performance-related objective level, and it must do so at an acceptable cost.

This approach can also be illustrated by a diagram showing the process followed. Seen this way, ecodesign is a design aid to reach an objective only through optimizing technique: Which solution, at a given cost, makes it possible to respond to a designated performance?

A project does more than design a shape, it designs an object.

The aim of the project management team as a whole is to deal with objects, and to do so on several levels. The planned construction is primarily a physical object represented by a shape. It is the result of a design that is in itself a complex process strongly influenced by the experience of the project managers.

However, this object is also the result of a combination of numerous elementary objects that make up a whole. This connection between the different levels of organization operates via a succession of critical evaluations. The outcome is itself dependent on the designers' personal requirement level.

The result of this two-fold determination is that the status of the object is not definitive, but on the contrary must be viewed as the result of a manipulation in space of physical elements. It is in this sense that a project designates the object of a choice.

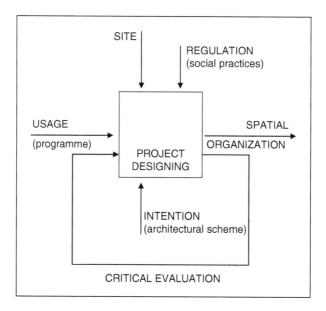

Figure 2.5 Suggested modelling of the project designing process of an architectural plan.

2.1.4 The nature of objects

Ecodesign presupposes that the project is continually optimized right from the start.

When it comes to the actual nature of the project design phase, ecodesign can shed new light. If the project puts a strong accent on spatial organization, this should be understood not so much as a personal decision as a response judged to be satisfactory with respect to the general contractor's expectations.

Starting from the rough plan stage, ecodesign can help show whether the process underway is likely to attain the expected results. It thus contributes to supporting the selected hypotheses and extending resolutions made in full knowledge of the facts.

This approach is in fact essential at this stage since it is the only point at which its effects wield the most weight. Increased efforts (investments in prior simulations) guarantee genuine project savings.

This observation is based on the industrial rule summed up by the following graph, which applies just as much to construction.

Projects characterize the objects to construct.

However, the way the architectural object is organized in space is not sufficient to define the project in its totality. It needs to be completed by a mode of organization. This term implies a capacity to set out the ways by which this precise situation can be accessed in a 3D reference frame.

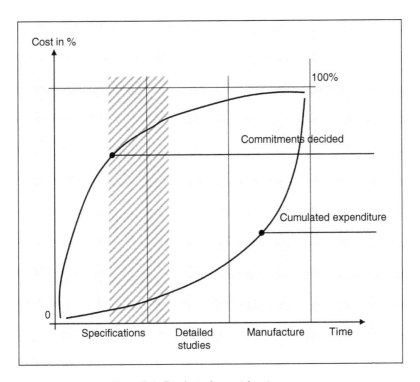

Figure 2.6 Ecodesign's special action zone.

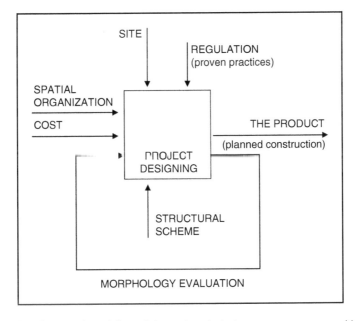

Figure 2.7 Suggested modelling of the project designing process at structural level.

This other dimension, which is a prerogative of technical project management, is translated by a mode of production that professionals designate as the transcription of the structural scheme. The form is transcribed by the positioning, material, texture or physical properties that must be respected. All of this information clearly characterizes the object, but its main objective is to set out the future construction conditions.

To some extent, project designing is a visualization of what the project will be like when it is handed over at the end of the project.

2.1.5 Dimensions conveyed by objects

Ecodesign anticipates all of the operating modes.

Ecodesign enriches this set-up by adding the notion of operation.

An ecodesign project cannot be restricted to a static dimension; it must anticipate how it will be used in the future. Ecodesign takes into account the choices made, such as orientation, neighbouring masks, and compactness, all of which are stable, long-term project data. However, it also combines them with usage scenarios that make projections over the long term.

This introduction of a time element into decisions is not neutral. It necessarily goes with a questioning of the financial dimension, which must thus distinguish the investment cost from the operating cost. In the long term, the financial aspect will be measured against the "holding cost", i.e. all of the costs induced by using the building.

Overall, ecodesign has a very significant lever effect on the conditions for putting together a project, since it leads to taking into consideration aspects that have been totally ignored up until now for very different reasons (e.g. denial of environmental issues, incapacity to measure, refusal to put into figures).

> *Project management work is subject to detailed 1973 regulations on "Engineering projects"*

The contract that mobilizes the project management vis-à-vis the contracting is often the result of an architectural competition.

The way practitioners are organized can be symbolized by the graph below. The variants only apply to the volume of interventions stipulated in the contract.

These variants are taken to be the supply of intellectual input that makes it possible to define a building in response to a programme. Note that this artistic property is inalienable in French law. However, they may be extended to include follow-up on the project by accompanying site work.

2.1.6 Pre-contractual form

Ecodesign covers all contractual configurations.

Ecodesign is above all an optimization approach. It therefore adapts to the organization of the project. On a contractual level, it does not introduce any specific features.

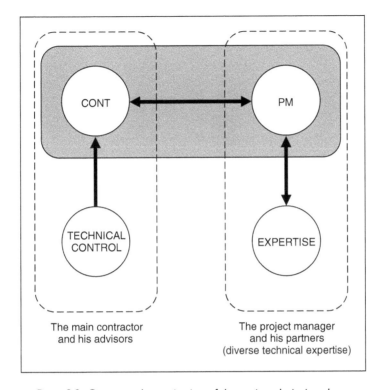

Figure 2.8 Contractual organization of the project designing phase.

However, it is clear that ecodesign is particularly easy to use and effective when it operates as part of a performance-based contract. It then corresponds to a "design with designated targets".

This organization scheme is the object of an industrial standard, XP. COD.

2.2 CONSIDERED PARAMETERS

The primary object of buildings is to provide useful surface areas for end users.

To successfully implement a project, the first condition is to determine the central object of this process.

The objective is to succeed in creating areas that correspond to the terms of the programme. These spaces will start off with the flooring, which creates the surfaces that will be available to the end users. This observation may appear self-evident, but it is nevertheless the key to achieving a pragmatic result, in other words one that is adapted to the expected usage.

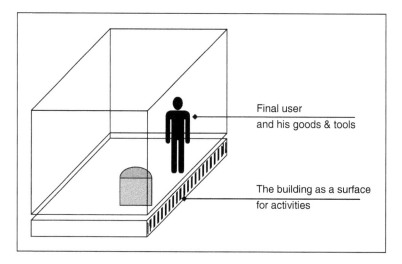

Figure 2.9 A construction as the creation of an "artificial" floor to carry out an activity.

This characterization of a project is defined by the production of a plan in which the total built surface area is measured in relation to the useful surface area only. Depending on how well a project design is carried out, it might result in additional circulation, or superfluous areas in line with the spatial choices made.

2.2.1 Creation of surfaces

However, structural choices are not neutral vis-à-vis the environment.

Ecodesign is directly interested in the development of surface areas, the optimization of which is translated by a lower volume of materials used.

However, two other parameters are taken into account: (1) the nature of these components and (2) the manner in which they are implemented.

Depending on the materials used (wood, steel or concrete), the impacts on the environment will be very different, as shown in the table below. The origin of these variations results from the scarcity of corresponding resources, but also the production processes, since each involves a semi-product combining natural elements in a manufacturing processes that often consumes significant amounts of energy.

In addition, on-site construction (the worksite) generates waste, the amount of which varies depending on the project's size. The issue at this level concerns the laying out, or in other words the relationship between the size of a component and that of the building, which to be optimal should be a multiple of the basic element.

Buildings constitute artificial "environments" adapted to end users' activities.

Table 2.1 Comparison of materials from two points of view: weight and CO_2 outcome. This dual-entry table provides a good illustration of the nature of ecodesign, which seeks an optimum between the quantity of matter mobilized and the resulting environmental impacts. However, this approach must take into account the material's performance on another register, i.e. mechanical. The choice is therefore not linear, but based on multiple criteria.

Material	Mass I m^3	CO_2
Iron	7,000 kg	+5,000 kg
Concrete	2,300 kg	+375 kg
Cement	1,600 kg	+2,500 kg
Hardwood	700 kg	−1,000 kg

The second element resulting from the designing phase is the building's envelope, which is initially apprehended in its aspect of volume and shape.

This sensitive approach, which should not be confused with façadism, is not simply concerned with the decorative aspect of the envelope.

In fact, it is a parameter of the architectural composition that is nevertheless translated by producing surfaces, or more exactly, facets, that mark the transition between the inside areas and the outside, which is subject to "natural elements".

The envelope thus contributes to sheltering users and participates in the building's very purpose, which is to protect from the weather.

It is primarily characterized by its built-up area (perimeter of the façade) and its opening percentage.

2.2.2 External delimitation of volumes

Envelopes insulate the inside atmosphere from the outside. They play a major role in energy saving. On this level, ecodesign favours the bioclimatic approach already written into the programme.

Ecodesign introduces a fairly different point of view regarding the envelope, i.e. its contribution to heating requirements. This is because the envelope behaves like a screen that on a physical level acts as an exchanger of warm inside flows and much colder outside flows, or vice versa depending on the season.

Some professionals qualify the envelope as a "thermal shield". Two characteristics need to be taken into account: (1) the nature of the material used and (2) the permeability to air resulting from the assembly of components in the sub-structure (opaque parts and openings).

Taking these elements into account involves an indispensable dialogue between the architectural designer and the technologist. The design can no longer be dissociated from its feasibility. In the English-speaking world, this crossed analysis is called "constructability". Ecodesign thus considers that, depending on the orientation, six analyses are required to define an optimal envelope.

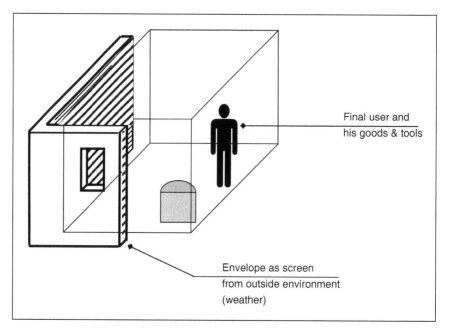

Figure 2.10   Buildings create a suitable atmosphere for activities.

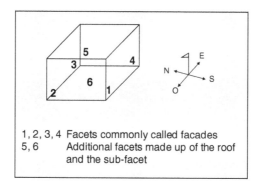

Figure 2.11   A building's envelope is viewed as a set of six treatments adapted to the orientation.

The available floor surfaces are divided into zones delimited by partition walls

The third element in a project is represented by what are commonly called partition walls.

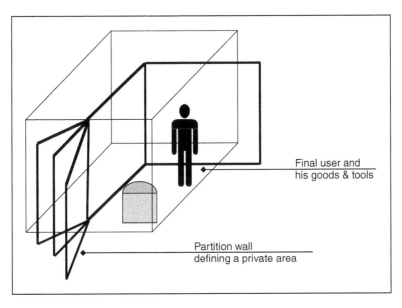

Final user and
his goods & tools

Partition wall
defining a private area

Figure 2.12 A building contributes to the intimacy of uses.

These partitions result from a need to allot a certain number of areas to different activities. This inevitably requires delimiting zones and separating them from the other areas, which are themselves divided between several users.

This division then creates a need for specific accesses, which will require using additional surface areas known as circulation and intermittent usage areas.

These space-dividing and distribution elements are themselves characterized by their lines, and the quality of a project is measured by their reduced number.

2.2.3   Internal delimitation of volumes

A decision to create partitions is directly affected by the changes anticipated over the building's lifespan.

In ecodesign, partition walls do not have any particular status. It is however worth making two complementary remarks.

By definition, partitions delimit homogeneous activity zones. They also serve to determine thermal regulation areas. However, these two divisions do not necessary fall in the same place, even though they may share some edges. When a component belongs to a zone, its purpose needs to be defined.

Partitions also need to be considered over the programme's entire life cycle. So that they can respond to any changes in the internal organization, they should preferably be designed using a dry-casting process that gives a degree of flexibility. This attitude may lead to the choice of an "open space" type structure.

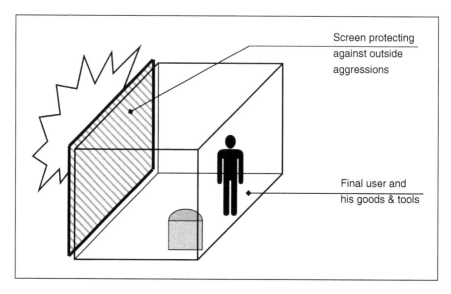

Figure 2.13 Buildings vis-à-vis personal integrity.

It is fairly important to realize that the question that underlies the definition of this operational sub-set is directly linked to the demands set out in the programme. Any variations in the partitions will depend on two objectives: (1) being able, for the acceptance of the completed work, to correspond as closely as possible to the end user's expectations, (2) enabling rehabilitation during the building's life cycle. Two distinct ideas emerge, whatever the scenario. The first of these is the contractor's obligation to stipulate the usage hypotheses that he intends to assume. The second is to make very clear the anticipated duration of usage. A technical decision will need to be made in full knowledge of the facts.

The selection of the materials will also depend on the programme's parameters. Even fixed partitions can be mobile, like the removable walls used in Japanese architecture. The important factor here is the reliability of the technology and its performance.

Buildings must permit secure use for each user.

Buildings are also places that protect users, which they do in several ways.

The first of these is from outside aggressions: resistance to wind, fire, earthquakes and humidity. In this case, buildings offer an adequately sized response to ensure security over a certain period of time. In general, this is done using proven calculation methods.

The second relates to voluntary aggressions that might threaten users' "tranquillity". The object here is to provide resistance to infractions or unwanted intrusions. Methods may be static (setting up barriers) or dynamic (alarm and override system).

2.2.4 Personal integrity

However, buildings fit into a broader context the health aspect of which needs to be preserved as a precautionary principle.

Ecodesign takes this "bodily" protection into account. However, an additional dimension comes into play here, i.e. the health aspect of the materials.

This term needs clarifying, since its field of application may raise some doubts. Ecodesign supposes that the project is perfectly ventilated (i.e. renewal of inside air). This prerequisite is an integral part of the functional unit. The idea here is to enlarge the perimeter of analysis and consider the external impacts generated by ventilation. This therefore involves public health. The objective is to avoid increasing risks to the community.

Ecodesign therefore involves choosing materials that do not release VOCs or potentially dangerous emissions during their life cycle. The corresponding indicator is still the subject of expert discussion to reach consensus. However, it is already possible to refer to information supplied by industrials (listing following protocol P01-10-01).

Buildings are subject to local bioclimatic treatment.

The project also considers the users' relationship with the outside environment of the site. Four factors are taken into consideration:

Landscape: This is essential for both users, who must benefit from the site, and for the community, which will need to accommodate a new built object.

Orientation: The building's position vis-à-vis the sun, along with dominant winds, should enable it to benefit from the gains without the disadvantages.

Services: Buildings cannot exist on their own, they should be devised to be in synergy with all of the neighbouring collective facilities (local shops, communication networks, cultural activities, etc.).

View: This is important for users in their relationship to the outside environment so as not to feel imprisoned or shut in.

Box 2.1: Protocol for measuring health impact in an LCA.

DALY INDICATOR:
Damage factors for human health are determined based on a list of 1,000 chemical substances as follows:

– Intake fraction (kg inhaled/kg emitted):
  Fate factor × Exposure factor;
– Effect factor (DALY/kg inhaled):
  Risk potential × gravity.

The indicator thus results from converting kg of equivalent substance into disability-adjusted life years (DALY).

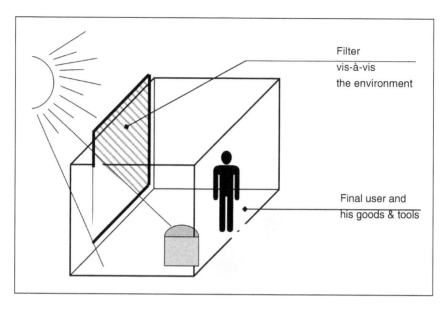

Figure 2.14 Buildings' relationship with the environment.

2.2.5   Integrating into the site

Buildings are also elements that constitute the world's environment, and eco profiles take this into account.

Based on the elements above, ecodesign extends the analysis to consider the project's contribution to global issues.

These are obviously included in the detail of the programme's declared performances. However, during the designing phase, these objectives can be reached in different ways, depending on the choices made by project management.

The withdrawal of raw materials is the result of consolidating the volumes of material used. All too frequently, the line of thought focuses on a single channel (e.g. concrete, steel, wood). Ecodesign aims to use materials where they are most advantageous. It favours a mixed approach and avoids structural monopolies.

Similarly, emissions should be viewed not just in terms of absolute value, but also in terms of existing collective facilities and their treatment capacity. Liquid emissions should not necessarily be recycled autonomously if pooling them at neighbourhood level results in a better yield. This is a good example of the advantages of making a precise analysis of local capacities.

Regarding energy consumption, the issue is very similar. The programme sets a consolidated level that authorizes all combinations of structural choices and types of equipment. The aim is obviously primarily to minimize energy requirements, but this can involve procuring supplies in different ways by mixing sources and taking advantage of local energy policies.

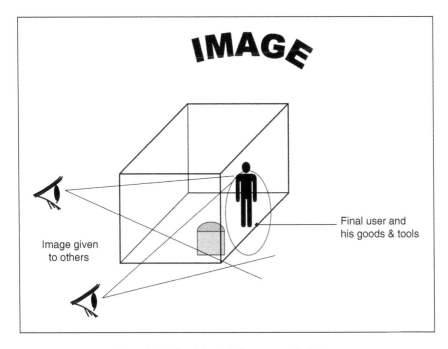

Figure 2.15 Semiological character of buildings.

> *Buildings are not neutral. Users experience them in very different ways.*

The relationship between users and buildings is complex. We can however distinguish two aspects of this relationship qualified as semiological.

The first component is how individuals feel about the built area. This reaction involves volume, shape, materials and colours. All of these parameters come together more or less harmoniously so that people take possession of a space to a lesser or greater extent.

This feeling is strongly related to the image given out to others. This is because a home also expresses a degree of personalization that is perceived by the rest of the community. As the saying goes, "Show me where you live and I'll tell you who you are".

2.2.6 Projecting an image

> *Ecodesign makes a major contribution in this area. It can be used as branding but may also remain discrete if it takes the form of more detailed treatment.*

Ecodesign is highly concerned by this perspective, which involves establishing whether it is capable of changing the image of the building industry.

In France, ecodesign is frequently associated with the notion of bioclimatic architecture, and either a specific material, wood, or the conspicuous display of some types of solar equipment (thermal or photovoltaic panels). This is acceptable if the results correspond to a target written into the operation programme. However, it is not a foregone conclusion, even though ecodesign is not totally neutral since it introduces new perspectives.

An analysis of the building's envelope is not restricted to the treatment of the façade. In fact, it involves reflecting on the five visible facets: the four sides dictated by the building's orientation, and the roof, which is often overlooked. In ecodesign, each of these surfaces requires a particular, observable response in line with its specific context (differentiation of gains and exterior agents).

The integration of certain structural elements will also bring more or less controlled impacts. These enter into two categories: elements that constitute semi-collective areas (e.g. balconies without thermal bridges) and those that contribute to producing energy (i.e. wind or solar power). Rather than constraints, all of these components should be viewed as enriching the spectrum of the architectural composition.

On a semiological level, ecodesign thus appears to offer a source of renewal and a way of avoiding the false values of façadism.

2.3 OBTAINED RESULTS

The way the project is organized in space is not just spatial distribution.

The first stage of project designing involves defining the position of the different activities in the space. This spatial organization, which is translated by the way volumes fit into each other, transcribes an architectural scheme that expresses an intention. It is very difficult to model because it is always a personal, often individual, practice.

This explains why the result generally puts the emphasis on connecting the surfaces rather than a simplified load transfer. In fact, creating superimposed surfaces implies seeking support on the ground. This is the natural law of gravity. However, the principle is all too often circumvented or even ignored due to current capacities to direct efforts in a very complicated manner that is fairly far removed from simple equilibrium. However, this approach comes at a price, which is the additional matter added to the bare minimum corresponding to the laws of physics.

2.3.1 Choosing architectonic principles

Ecodesign involves reflecting on the best structural scheme for the project. The choice starts with the best use of material.

Ecodesign means turning this practice round and thinking about how efforts can be focused to simplify the construction.

Interestingly, no single solution exists. On the contrary, several principles are possible. In addition, each of these can be undertaken using different technical channels, as shown in the figure below.

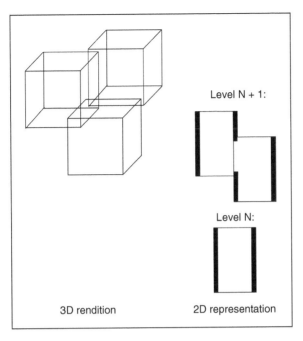

Figure 2.16 Spatial organization of areas represented in graphic 2D transcription.

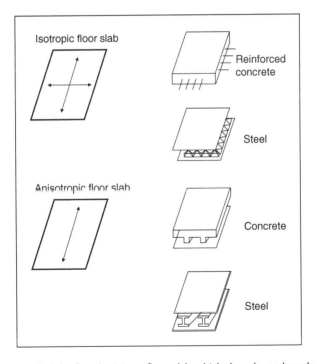

Figure 2.17 The two principles for obtaining a floor slab, which then depend on the materials used.

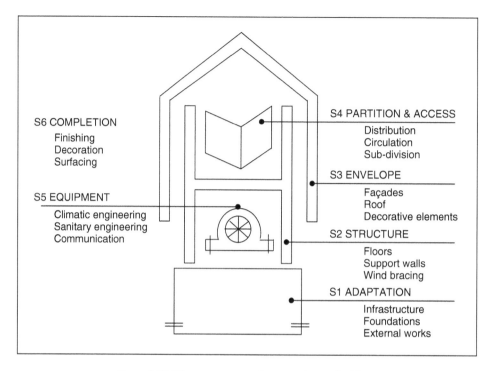

Figure 2.18 The components that constitute a building.

In fact, this approach concerns all of the components that make up a building and, taken together, the principles make a useful set of vocabulary for managing the project.

An ecodesign approach puts the emphasis on the close interweaving of the architectural and structural principles. The term used to describe this discipline is "architectonic", and its origin is as old as architecture.

> *A project takes shape through specialized elements that make up the functional subsets.*

The second stage of the designing involves defining each of the parts that physically represent the project. This analysis relates to generic sub-sets since they can be found in all architectural configurations.

A list can be drawn up that includes six families, qualified as functional subsets because they specifically correspond to some of the functions expected by the end user.

The project therefore creates spaces that it makes concrete through choices relating to each subset. This work, which may come across as too methodological, nevertheless conditions the operation's feasibility. What is clear is that it is not possible to do without any one of them.

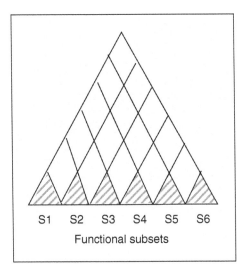

S1   S2   S3   S4   S5   S6

Functional subsets

Figure 2.19 The systemic vision of a building: elements interacting through their interfacing.

2.3.2   How components link together

Buildings are not simply an addition of elements, they are interacting wholes, which is what distinguishes an ecodesign.

Ecodesign adds a complementary point of view to the above approach, and that is a systemic vision.

Although each functional subset is sufficient on its own, the project itself only results from the overall coherence of these components. A building is a whole that behaves as a whole. Expected performances will be significantly diminished by any solutions to create continuity between elements.

This observation, which is as old as the art of building, has nevertheless taken on greater significance with the energy issue, which has highlighted the importance of interfaces: linear thermal bridges are now a crucial part of thermal balance sheets. Yet with the increase in thermal performances, the question of airtightness becomes unavoidable and by nature can only be dealt with through interfaces.

The precise determination of each structural element is essential in characterizing a project.

After selecting an architectonic principle and envisaging the different parts of a building, the project design will focus on defining each component.

The rough plan stage has already been achieved at this stage, and with the characterization of components, a new phase begins: that of the detailed preliminary design. This progression is explained by the French practice of using reinforced concrete: because it is a malleable material, a number of questions relating to the construction no longer need to be posed.

This detailed preliminary design takes the form of a drawing showing the building's geometry, which comes under the responsibility of the technical project management. In general, no difference is made between the structure and the envelope, which are cast in one piece. This so-called "wet-cast" process is now so common that it has become standard, often carried out as a routine practice that overlooks the need for careful building planning.

This easy use is specific to concrete. It would be impossible, for example, for a metal-based structure, which would involve bars used to carry the weight in a much more linear way. In this case, the structural process is discernable and visually conveys the designer's skill. These remarks show the importance of the choice of a building principle, regarding both the shape and rigorous judgement.

2.3.3   Choice of building principles

This characterization results from choosing between a range of possibilities that can direct the environmental profile in the construction phase.

In ecodesign, the process is very different. Each component is thought through much more carefully and analyzed to determine its structural principle.

To illustrate this approach, we can take the example of a heavy envelope. The first remark worth making is that a heavy façade is not necessarily structural, and that it is primarily a solution filled with heavy materials. To fulfil its thermal screen role, three options exist:

– Internal insulation can be used to supplement the "mechanical" shield, but a linear bridge treatment should be envisaged;
– External insulation can be rolled down in one piece, which solves the thermal bridge issue;
– A material that is both insulating and resistant, i.e. embedded insulation, fulfils both of these functions at once.

Projects should anticipate potential difficulties on the production site.

On reflection, the project-designing phase precedes the construction, and in this sense it is a good time to think about the building methods. In English-speaking countries, this is called constructability, which means identifying obstacles before the project is actually built to ensure that it is achievable.

This work is not necessarily done in the rendering of the project because in traditional processes the call for tender leads to the necessary adjustments.

The technical project management team, which might normally be responsible for this phase, most often simply approves the sizing, but never produces execution plans. The reason for this may lie in the fairly low remuneration paid out in line with public engineering scales for this task.

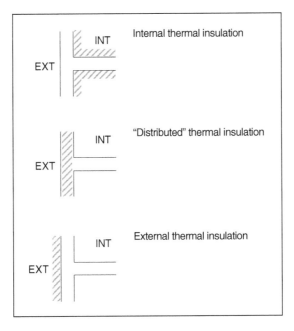

Figure 2.20 The principle three families of heat-efficient envelopes.

2.3.4 Constructability

When ecodesign focuses on the interfaces, it contributes to this anticipation.

In ecodesign, however, this stage is crucial. The reason for this is that it conditions all of the savings made by the project because its objective is to make things possible. This precautionary approach is thus a guarantee of success.

Constructability involves making a dual analysis of each element. In a first stage, all of the constraints imposed by the rest of the components must be listed so that the component in question can be adjusted to these imperatives. The second stage involves listing the specific characteristics of the component, which will need to be verified before putting in place the rest of the building.

This approach can be illustrated in a matrix that shows how the interfaces between the components and the building are managed. Constructability plays a big role in managing the quality of the project and also by anticipation in managing the on-site production.

A project is often accompanied by a simple instruction on its components, i.e. a definition of the means to be used.

Along with providing a set of graphic information on the building, the project management team must also provide instructions regarding the characterization of the means selected.

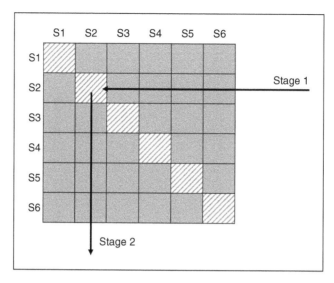

Figure 2.21 The two stages of a constructability analysis: Stage 1: Definitions of the interface; Stage 2: Conditions for approval & support.

Traditionally, this task is carried out by the technical project managers, who start by producing a justification of the building's size. The work is done with reference to documents or rules that have an official character in as much as they are cross-cutting modes devised to homogenize practices and avoid counter-performances. All calculation codes thus implicitly constitute the validation of a performance level.

Along with this justification of conformity, all projects are subject to the French general and specific technical specifications (respectively CCTG and CCTP). These written documents recommend practices that include the results of general feedback on experiences (references to Code of Practice, etc.). In fact, this entire framework indicates applying tried and tested know-how, which is judged necessary to deliver a building in working order.

Taken together, these data not only format the project that is to be delivered, they also constitute an approach that is essentially focused on geometry and that considers the rest of the performances as tacit since they are in line with regulations.

2.3.5   Level of performances

Ecodesign involves a specification approach, i.e. one that designates the performances of each part of the building.

In ecodesign, the approach is different because it does not simply involve making sure that the project is in conformity. The objective here is to attain a higher level of performance. It is a design with designated objectives.

	Principle	Performances	Continuity
Area	Orthotropic room	Size Overload Deformation	Connection to envelope
Atmosphere	Mass (acoustic) thermal break	Acoustic reduction level Ko	Treatment of linear bridges
Protection	M1 material	Respect of fire regulations Seismic regulations	Treatment of technical floor openings
Relationships	Vertical circulation passage	Header beam performance	Load carrying
B&O	Network incorporation	Diameter to incorporate	Connection reservation
Site	Projection towards outside	Size Overload Tightness	Connection to surface
Semiology	Bare lower siding	Geometry of joints Surface porosity	

Figure 2.22 Example of listing of performances for flooring.

Since these objectives are detailed in the functional programme, the project design phase will also be based on these values and will apply them to each structural element. Thus, each response "object" will be accompanied by information structured according to a matrix approach.

The constituents set out in the project all contribute to accomplishing the operational performances (the seven usage functions) because: (1) they result from a choice of principle, (2) they present specific elementary performances that correspond and (3) their continuity with the rest of the building is stipulated.

These performances are accessible in the form of tables.

> *The decisions are written into specifications considered as a reference document that substitutes the initial programme.*

As indicated previously, justifying the choices made traditionally involves verifying whether buildings conform to the various regulatory codes. This is therefore less an explanation of the choice than a simple validation.

2.3.6 Justifying choices

In ecodesign, choices are supported by a set of simulations that antici-
pate the future behaviour of the building in a broad range of situations
so as to respond to the programming demands.

In ecodesign, the justification is not restricted to validating in line with regulations, it also entails illustrating the building's behaviour according to different points of view. This is done using simulations. This term implies the capacity to visualize the long-term behaviour of the whole building.

2.4 TRANSFER MODALITIES

The plan documents are at the heart of the project. They transcribe
how the operation will definitively take shape.

The project is ultimately translated by a set of graphic documents that will enable the building to be produced.

It is however indispensible to look at the purpose of these elements. These are essentially viewed as the formal translation of an architectural project, i.e. they formalize the geometric shape that the project managers have decided on and provide the measurements of the constructions to be built.

However, this translation is codified by a tradition that is very different from industrial practices in that it imposes absolute measurements rather than indicating manufacturing tolerances.

These documents thus in no way constitute construction documents that can be directly used by future practitioners.

2.4.1 Architectural plans

Ecodesign favours detailed plans that provide more information than
just geometry.

In ecodesign, the objective is to possess annotated plans, i.e. including complementary information that makes it easier to define the buildings to be built.

These plans must be used directly to ensure the last simulations, which implies knowing the exact nature of each material to inform the calculation data.

In particular, the performance level reflected by ecodesign requires a great deal more detailed designs. These will relate to the definition of interfaces, which is crucial to successfully tackle the question of permeability (aeraulics, acoustics, etc.).

These practices are more common in the Anglo Saxon sphere, but they are making ground in France.

All project documents include special technical terms and conditions
(CCTP) and general technical terms and conditions (CCTG).

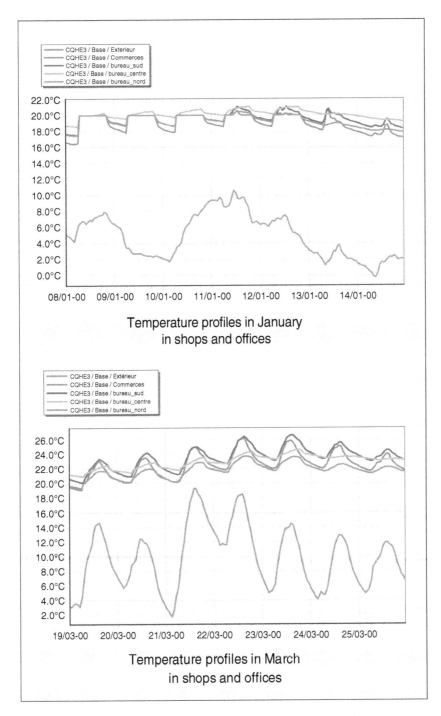

Figure 2.23 Example of results from a dynamic thermal simulation.

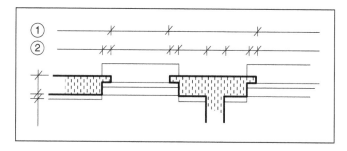

Figure 2.24 Sizing principles for traditional plans: I – cumulated sides, 2 – detailed sides.

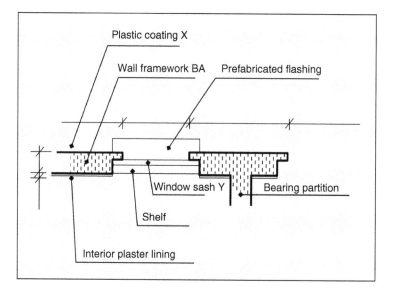

Figure 2.25 Example of a detailed plan.

Traditionally, project plans are accompanied by numerous written documents that consolidate useful information for the actual production. These come under two major families.

The first of these provides a more precise description of the work to be done. Note, however, that this document is primarily organized according to production specialities and is broken down into building trades. This is the principle that contributes to the allotment.

The second family relates to good practices, i.e. the rules of the art codified by the different unified technical documents. These elements determine the conditions of performance so as to limit any faults. The focus here is on ensuring that the resources are used appropriately.

Taken together, these documents create a voluminous package considered as the sum of precautionary measures. It is not always easy to match them with the plan documents.

In all cases, these administrative papers are accompanied by a CCAG (general technical terms and conditions) and a provisional building timetable.

2.4.2 The actual documents

Ecodesign involves a thorough definition of each structural element of the future building.

In ecodesign, the written documents are no longer viewed as instructions, but as specifications. Rather than defining the means, they set out the level of performance to reach.

In this case, comprehensive plans are accompanied by details on the expected performance of each component of the future building. This is called a "performance approach".

It defines the geometry of the buildings (size conditions) and all contributions relating to sustainability (resistance, lifespan), energy (consumption levels, permeability, etc.) and the environment (nature of accepted impacts)

This package requires close work between the architectural and technical project managers. It is the outcome of a meticulous analysis that envisages the project in operation rather than applying routine solutions.

Building projects cannot be isolated from their financial aspect. Traditionally, calls for tender are considered as "market approval".

The validation principle for all projects obviously involves respecting the budget allocated to the entire operation.

The most common approach is to carry out a call for tender. This corresponds to a concern to bring genuine competition into play between bidders. Over and beyond this, however, the objective is often to ensure that the project is genuinely feasible in the expected conditions.

This validation is thus external to the project and carried out at the end of the phase. As a result, any corrective measures come late and cannot take the form of a change in the services offered.

Specialized intervention from an economist could be envisaged at an earlier stage. However, this kind of evaluation is done on the basis of ratios that are not necessarily adapted to the project's specific features. In addition, the remuneration of practitioners in this designing phase is often based on a rate (percentage) of the turnover, which does not encourage looking for genuine project savings. This is another reason for making use of approval a posteriori from a call for tender.

2.4.3 Validation conditions

Ecodesign does not work ex post, it is put together ex ante, anticipating the causes of surcharges.

Ecodesign is based on the opposite process, which involves anticipation and is the result of an approach in terms of constructability.

The early reflection on construction difficulties cannot be reduced to the future technical costs. This detailed analysis is translated by the comprehensive definition of interfaces contained in the graphic package. This kind of approach is demanding, but it is the only way of obtaining reliable final results. It has the effect of reducing the risks of production and thus diminishing the hazards that producers necessarily allow for in a standard approach.

Overall, it illustrates the fact that "a problem clearly set out is already partially solved".

Based on this organized preparation, it is possible to establish a detailed bill of quantities and put a figure on the planned building. Adjustments are possible if it the amount calculated does not correspond to the budget allocated in the programming phase.

> *The project usually ends up with a contract that is a best efforts obligation.*

The project is transmitted to the practitioners working on production via the call for tender documents, which can be used to select a team. This transaction operates using a contract the nature of which is worth analyzing.

Generally, the object defined is a response to an order. Thus, the contract deals with best efforts obligations.

Legally, this consists in respecting all of the information contained in the graphic package and in the two sets of technical terms and conditions (CCTP and CCTG).

2.4.4 Post-contractual form

> *Ecodesign is based on obtaining performances. It therefore accompanies all results-based contracts.*

In ecodesign, the terms of exchange written into the contract are of a different nature.

Transactions only relate to obtaining a reasoned result, without imposing a prerequisite solution. In other words, a contract is only considered as fulfilled when results can be verified in real size at the end of the production.

As a result, the contract includes a new element, which is the verification protocol linked to each of the "performance" specifications. Implicitly, the very fact of establishing a quantitative performance level implies that it is possible to measure it.

> *Project construction is entrusted to the lowest offer after a call for tender.*

The result of the project design phase is the designation of a team responsible for production. This team's remuneration is linked to the project.

All too often, this payment is based on a notion of the lowest bidder. This involves choosing the least expensive offer resulting from the call for tender.

This procedure, which originally responded to a concern for competitiveness, is gradually being brought into question. The principle of this mechanism is in fact based on looking for the best unit prices for totally defined objects, with a risk of drifting towards lower prices to the detriment of work of quality.

2.4.5 Payment

> *In taking a best offer approach, projects fully associate those who will be producing them.*

Ecodesign that is based on performance leads to a new form of payment, which some describe as "best offer".

The objective, based on the performances put forward, is to obtain a response that reaches the required level, and sometimes more, due to freedom in choosing the means.

This capacity to mobilize innovation should be put into practice with some caution since the idea is not to play "Russian roulette". The target results should remain accessible in a context of available or probable technical expertise.

2.4.6 Synopsis

A project design must end up with a result of which the future operation will be totally in line with the expectations expressed in the early programming phase.

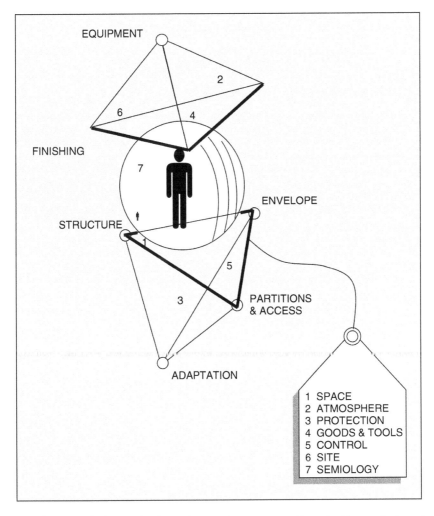

Figure 2.26 Symbolically, the result of a project can be represented by a double tetrahedron showing the different operating functions expected by the user.

2.5 OVERVIEW DIAGRAM

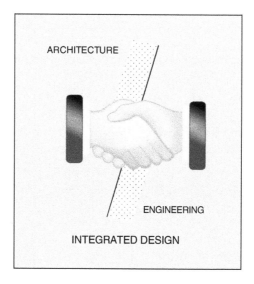

Figure 2.27 In ecodesign, projects are facilitated by close cooperation between two disciplines, an idea already put forward by Le Corbusier. This co-development process guarantees efficiency. In addition, it reinforces the fact that there are no ideal solutions, only solutions that can be considered satisfactory for specific contexts.

A project that is always put in context, despite its specific character, stems from a logical, thought-out organization that takes the form of a detailed, gradual work process and takes previous experience into account.

Chapter 3

Ecodesign in the production phase

By force of circumstance, construction companies have become the building pivot of projects.

3.1 MOBILIZING THE ACTORS

The main actor in this new project phase is the construction company. The builders make the project concrete so that it will be available to future users.

The construction company's role is focused on the sole aim of taking the project to its "successful completion". To do this, it uses skills and resources along the lines of the diagram below, whatever the contractual framework of its intervention. Its mission can be qualified as logistical support for the production team. This activity can be broken down into four main phases, whatever the type of building. None of these phases can be overlooked. Although they can take diverse forms, they essentially involve: (1) mobilizing resources, (2) using them in the construction, (3) making them into a building, and (4) controlling the proceedings.

These four components depend on specific know-how that can be enriched by feedback on experience.

3.1.1 The construction company's role

Ecodesign supposes that construction companies act responsibly and take particular care in managing their worksites.

Ecodesign at this stage should be understood as a re-reading of each of the phases from an environmental point of view.

This involves always using resources and organization methods that generate the least impact on the environment, both close to the worksite and further away, taking global issues into account.

Ecodesign therefore involves working on the procedures implemented to fulfil each of the sequential tasks. The efficiency of an ecodesign is thus measured not so much by the gains obtained as by an approach illustrating the desire to behave as a responsible practitioner aware of the need to participate in a collective installation.

This observation is necessary to resolve a misunderstanding. Ecodesign has more impact in the project design phase, i.e. when the essential choices are made. During

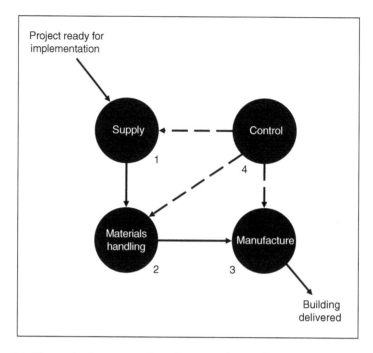

Figure 3.1 The production process from the point of view of the construction company.

production, progress can still be made, but to a much lesser extent. In a traditional building, the relative weight of the production including manufacturing the materials does not exceed 10% of all of the impacts generated over its life cycle. This observation is significantly modified with low-energy buildings, in which it can come close to 30% while remaining fairly low in absolute value.

Nevertheless, thanks to its capacity to modify, which is all the more likely when the initial technical specifications are weak, a company can make full use of ecodesign to compare and highlight its work.

> *One of the main characteristics of production is the multiplicity of practitioners.*

Up to now, we have considered companies in the form of a general contractor.

However, even in this case, the contractor will not be a single entity, but a much higher number of entities known as "sub-contractors". The same is true when divided into several contracts.

Beyond the contractual context, a distinction can be made between own production and sub-contracted production. This leads to introducing a new task, which is that of coordinating the various "building trades". This coordination should make it possible to start off the construction with a joint approach by which each practitioner can work in appropriate conditions and make an active contribution to producing the building. In other words, the aim is to waste as little time, energy and raw materials

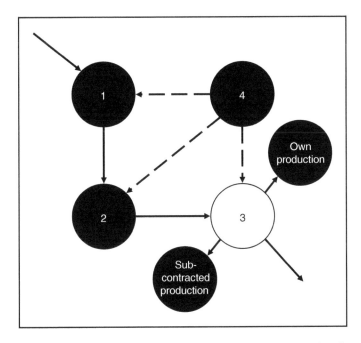

Figure 3.2 The manufacture can be integrated or outsourced, which will have the effect of increasing the number of companies involved.

as possible. This concern for saving resources is generally understood, no doubt too narrowly, in terms of seeking productivity.

3.1.2 The sub-contractors' role

This joint activity represents a risk to taking an ecodesign approach.

Once again, ecodesign does not change standard practices, except for in one area.

Outsourcing certain tasks does not mean overlooking the environmental impacts. In fact, the contractual configuration should necessarily be accompanied by two complementary actions:

1 raising awareness at the start of the works, and
2 tracing of the procedures used.

This implies much tighter monitoring, no longer restricted to just a responsibility report, but involving paying much closer attention to the means of implementation, which should conform to the hypotheses selected to establish provisional LCAs in the project design phase.

The share of a construction's turnover that goes to industrials is on an upward trend.

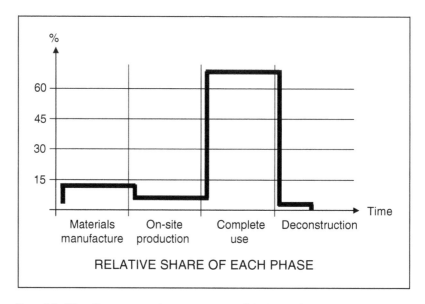

Figure 3.3 "Good" on-site production in terms of the level of environmental impacts.

A worksite's activity obviously depends on the workmen employed on the site. However, the materials and components that they manipulate have been manufactured by other workers at an earlier stage. Most of the resources used come from the industrial sector.

Industrials in the building sector fit into two main categories: producers of half-finished products and suppliers of components. The former provide items that are put together on site (e.g. ready-to-use concrete, metal work, insulation panels, etc.). The latter provide complex items that are integrated into the building (e.g. technical equipment, operating sub-units, etc.).

This distinction is debateable, but the main emphasis here is on the treatment carried out on site. In the first case, the transformation involves intermediate states before its definitive contribution to the building (e.g. preparing casing for a reinforced concrete wall). In the second case, the main task is to assemble equipment on site (e.g. a lift fitted into its cage).

3.1.3 The industrials' role

> *The ecodesign of the construction phase depends on the ecodesign of the different components.*

In ecodesign, each of the families of industrial suppliers is treated in a different way.

For half-finished products, suppliers must provide technical information sheets defining the environmental impacts resulting from manufacture and delivery of their

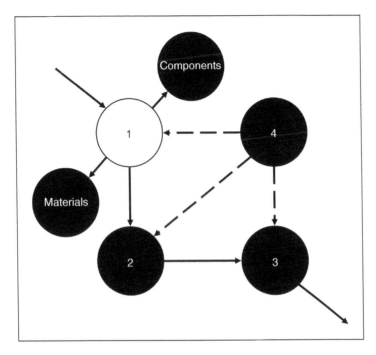

Figure 3.4 Supplies can be divided into materials and half-finished products to be assembled.

contribution. In France, the format of this information is defined in the norm P01-10-01.

For "ready-to-use" equipment, the process is similar, except that more information needs to be collected and consolidated. The following diagram sets out the procedure that is progressively being put into place in each industrial sector.

> *Buildings need to be linked up to the different networks existing in a location.*

The object manufacturing phase mainly takes place in the location in which the future building is to be constructed. Professionals tend to call this the "worksite". This illustrates another characteristic of this stage, which is the specific local context. For practitioners, this is translated by the specific role of distributors.

This term includes the suppliers of various flows needed to operate the building (i.e. energy, water, sanitation) as well as operators of communication networks. These inputs contribute to making the building viable and the installations operable.

Depending on the location and the measures put in place by the municipality, these distributors will condition the construction's timetable by designating the points and means of connection.

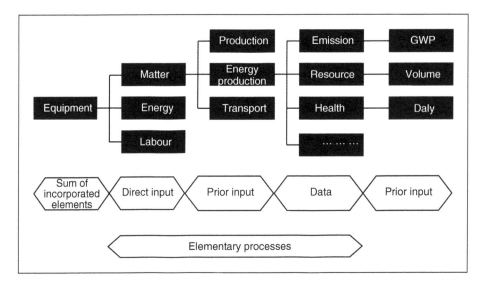

Figure 3.5 The problem of incorporating equipment into a building depends on the number of items involved.

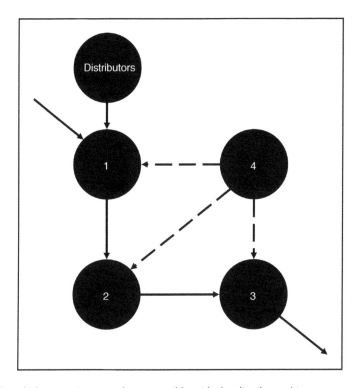

Figure 3.6 The whole operation must be compatible with the distributorship contracts that apply to the site.

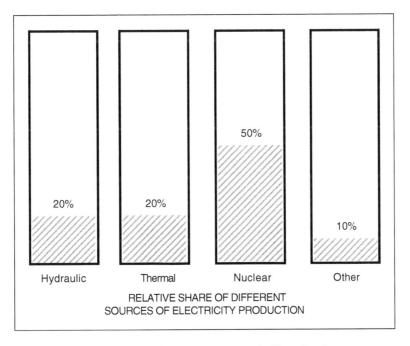

Figure 3.7 Example of an energy mix supplied by a distributor.

3.1.4 The distributors' role

The nature of available flows strongly affects the environmental impacts.

In ecodesign, the significance of these actors is high. They determine the supply profiles that will influence the level of environmental impacts linked to the operating phase.

During the production phase, it is necessary to verify whether the supplies agreed by the distributors match the hypotheses selected during the project design phase.

As a precautionary measure, the entire production is subject to an outside audit.

The technical controller's intervention in the production phase can be explained by two motives mainly linked to the notion of process management.

This control carried out by an organization commissioned by the contractor corresponds in quality management terms to what is known as outside control.

It involves making sure that the work done by both the project managers and the contracted companies conforms to proven practices. As this task is undertaken by a body that is independent from the other two entities, it brings an additional guarantee of the operation's successful completion.

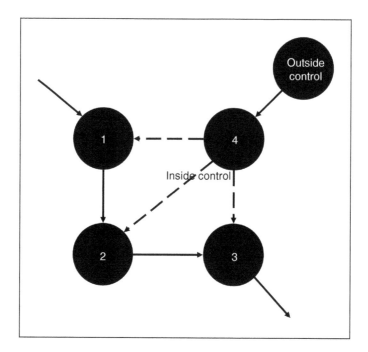

Figure 3.8 How the two control levels fit together in on-site production.

This intervention can also be explained by the French insurance system, which requires that the project be audited by an external expert. This precautionary principle takes the form of an examination of the working plans and corresponding calculation notes that precedes any implementation.

The technical controller thus creates a bridge between the project management and the contracting companies.

3.1.5 Technical control

Since ecodesign involves performances, its implementation also needs controlling.

In ecodesign, contrary to common belief, the technical controller does not play a greater role.

The only task that requires additional control is verifying the flows, which result from breaking down the installation into elementary installations, and their quantification in volume or mass.

Ecodesign is based on this bill of quantities and involves a conversion following pre-established matrices, which is then subject to specialist agreement based on approved common values. These databases must be displayed with each calculation but do not require specific verification.

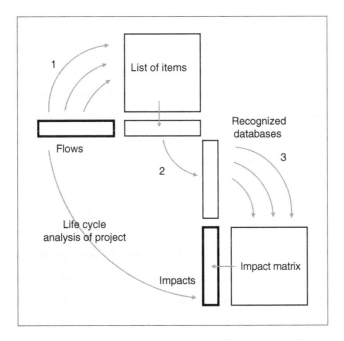

Figure 3.9 The principle of an LCA evaluation of a building: a series of matrix products.

The transparency of the impact calculation primarily relates to the bill of quantities, which needs to be approved by a third party.

> *Contractually, production sites result from a dialogue between three stakeholders (general contractors, project managers & building companies) organized by two contracts.*

Traditionally, production practices might seem to be managed solely through project management. Contractually, this is not the case.

In reality, it is the relationship between the contractor and the building companies that is subject to a contract. This explains why the company's responsibility is often reduced to simply respecting the means described in the clauses of this transaction, defined by the project manager.

The contractual form is thus defined as the transitiveness of responsibilities; at this stage the project management is only responsible for evaluating the work done by the building companies.

3.1.6   Post-contractual form

> *Ecodesign may be applied only to the production sequence in the building work contract. Yet it is most efficient when it is part of an integrated management system.*

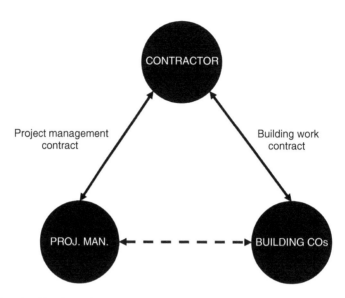

Figure 3.10   On a legal level, production depends on the direct relationship between the general contractor and the building companies. The project manager only follows up on the contract.

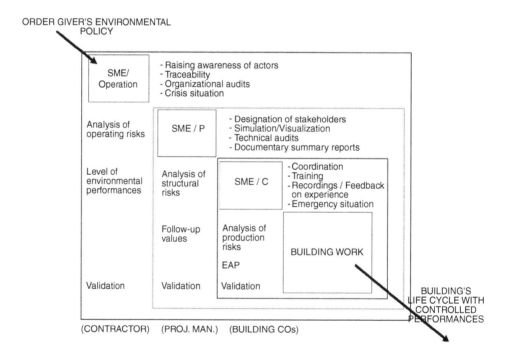

Figure 3.11   Production should be seen as the achievement of joint work that becomes more refined with the progressive contribution of each practitioner.

Table 3.1 List of building site stakeholders.

Neighbouring population	– Shopkeepers working close by who could be affected by the worksite – Residents who look out over the worksite and are likely to be exposed to nuisance
Local associations	– Pressure groups likely to appeal against the project – Residents' community associations
Local authorities	– Local police station for traffic and security issues – Municipal services for cleaning roads and collecting household refuse.

When ecodesign is based on voluntary commitment, it fits more deliberately into an integrated ISO 14000-type approach.

In doing so, contractually, it corresponds to an interlocking organization that attempts to optimize each contribution. The production phase in particular is necessarily the outcome of all previous action undertaken with the same concern to limit impacts.

This interlinking fits in perfectly with a "performance-based" contract. It is primarily based on a measurement concept, in other words, it puts forward objectives and verifies whether they have actually been reached by quantifying the built object's performances.

3.2 CONSIDERED PARAMETERS

Construction temporarily mobilizes a great number of practitioners.

Each project is specific in two different ways. The first of these is that professional workers are temporary because they are brought together with a common, but single, goal. In addition, the context of the operation is also specific.

The first condition for launching the production phase is to identify all local stakeholders and in particular those located close to the project who are not contractually linked with the production team.

3.2.1 The stakeholder spectrum

Ecodesign supposes managed behaviour that results from corporate environment certification.

Ecodesign only relates to subjects that meet with the rules of so-called "good practice".

This expression involves taking an approach that meets the requirements of an environmentally managed worksite. In other words, in line with ISO 14000-type specifications, the aim of which is to minimize nuisance generated by the production activity (e.g. noise, dust, traffic disruption).

On this basis, ecodesign is more particularly focused on the flows generated by the construction independently from those at its origin.

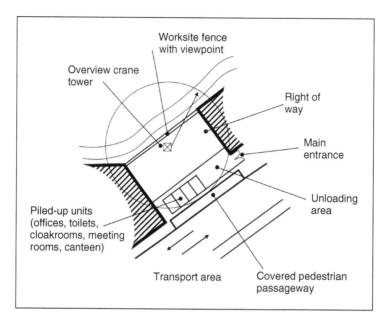

Figure 3.12 Example of a worksite set-up in an urban location.

An increase in impacts during the building phase may be acceptable if it leads to a greater reduction in the later phases.

> *The production site mobilizes its own resources that, even when temporary, are more or less efficient.*

The production site is a built unit itself. It includes its own buildings, roads and storage areas and communicates with the rest of the town.

The organization of the base is a project in itself that involves a block plan, structured flows of merchandise, general protection (fencing) and often a viewpoint that enables passers by to observe the work underway.

All of this preparation means that reception and production can be managed in the best conditions while providing meeting areas to complete administrative tasks. It also provides indispensible means for handling materials and distributing all of the work posts.

3.2.2 Organization of the site

> *Ecodesign must be applied to each component of the "base camp".*

At this stage, ecodesign finds a new field of application, which is linked to that of the built object. The site is a base camp that can be defined in ecodesign because it is a constructed object itself.

Figure 3.13 A surveyor is a historic practitioner. (Source: Matthias Kabel)

In particular, the prefabricated units on the worksite are traditionally high-energy consumers. This is because they are considered as temporary constructions and so do not come under common building rules.

In ecodesign, prefabricated units represent almost 60% of a worksite's electricity consumption. It is therefore possible to make savings if the same performance criteria are applied as to the long-term building. Solutions exist, and even though they entail making an initial effort, their additional cost is rapidly covered by the operating savings made (amortization possible over 5 years). These so-called efficient units generally have "GTB" (technical building management) certification, which means they can be adjusted to usage (pre-heating on a timer, presence sensor, etc.). Their consumption is reduced by over 70%.

This effort also contributes to improving working conditions for all of the staff working on the site.

The surveyor carries out the act that inaugurates the production.

The act that inaugurates the construction is the building's implantation on the site. This means that the geometry can be definitively fixed, along with its orientation vis-à-vis the built environment and its position.

3.2.3 The building's siting

Ecodesign uses standard measurements.

Ecodesign is not concerned with this issue.

However, the question of the building's geometry brings a reminder that the performance level put forward at the start of the programme needs to be set out, stipulating the surface area unit used. The following table indicates the different possibilities in French practices.

Production is closely linked to the nature of the subsoil.

One of the most common difficulties when launching the production of a building is a lack of precision in the nature of the subsoil. This information conditions the size of the foundations.

In general, additional surveys are carried out when the site is taken on, and are then used to make a definitive calculation of the foundations.

Another prerequisite is to confirm whether any network routes may be introduced into the site's right of way. If necessary, they may need to be moved.

3.2.4 Geotechnical prerequisite

The volume of material mobilized may end up modified.

This adjustment interferes with ecodesign, in as much as the quantities actually used are different from those in the project's hypothesis.

This proves once again how important it is to collect information right from the programming phase, given that the geological nature of the soil is exogenous to any project.

Box 3.1: The different units used to measure a building's surface area.

SHOB:	Gross floor area
	The sum of the total horizontal areas of the several floors of the building (not including floor openings)
SHON:	Net floor area
	SHOB minus the thickness of the walls and the surface areas of machinery compartments
SH:	Habitable area
	SHON minus the balconies, common loggias and passageways
SU:	Useful area
	SHON plus half of the surface area of the annexes

As an illustration, $1.4\,\text{m}^2$ SHON $= 1\,\text{m}^2$ SU

Wall thickness is around 5% SHON in standard buildings.

Source: Code la construction et de l'habitation

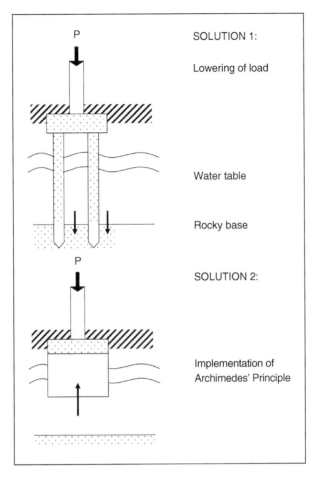

Figure 3.14 Example of structural principles for load spreading: I – Deep foundations, 2 – Compensation in a specific subsoil context.

Production takes place as the implementation plans are validated.

The production supposes that each part of the installation has been defined, geometrically (size), spatially (3D location) and physically (nature of performances). All of this information is included in a set of plans that are used as a reference throughout the worksite operations.

This plan package is supplied to the construction companies and is progressively detailed to enable precise implementation. However, this process requires a series of validations to reach total harmony between all points of view (formal plan of architectural conformity, verification of sustainability, adjustment with other structural elements).

When all of this information is broken down and examined from an implementation point of view, the building is considered as constructible.

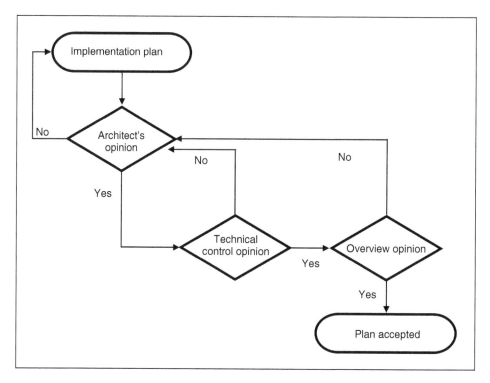

Figure 3.15 Approval route for acceptance of a plan.

3.2.5 Approval of the plans

These are translated by a definitive volume of resources used.

Ecodesign is primarily interested in the detailed bill of quantities resulting from defining the objects to be built.

Normally, this information can be cross-checked with the priced bill of quantities (DQE in French for "devis quantitative estimatif") that is handed in with the tender when the work is contracted.

However, a verification based on the operating plans is indispensible to support the survey statements.

A worksite involves numerous prerequisite authorizations.

To operate fully, a worksite must attain a certain number of authorizations that partly depend on the organization mode selected for production.

3.2.6 Authorizations

Ecodesign is not involved in this stage.

Table 3.2 List of worksite authorization requests.

Connections	– Supply: Water, gas, electricity, communication – Waste: Water valves
Equipment	– Crane set-up and oversail authorization – Worksite fencing – Worksite sign display – Storage of inflammable materials
Labour	– Declaration of staff – Exemption from fixed working hours
Traffic	– Road closure – Modification of pavement (pedestrian passage) – Traffic signal displacement (lights and ground signals) – Bus stop displacement
Flora	– Protection of existing trees

Table 3.3 Tasks organized by trade.

Rough work	Earthwork Foundations Structure Roof framing Exterior joinery Sealing
Light work	Lining Partition work Electricity (high power, low power) Heating/Ventilation Lifts Coatings Paint Metal work

3.3   OBTAINED RESULTS

Production is frequently organized on an allocation basis.

The key aim of this phase is to determine the different steps to be made and provide details of each of the elementary operations that constitute them.

This exercise is carried out in line with two rules, i.e. experience acquired in similar previous situations, and technical knowledge associated with the installation considered.

This observation may seem obvious, but it nevertheless follows standard practice by which this period is above all a continuation of gestures acquired on the field. This can partly be explained by the repetition of building techniques that are very similar from one worksite to another. We might even talk of a "Taylorization" of construction, although this would only apply to the definition of activities to be carried out, since the precise quantification of efforts depends on each situation.

The naming of tasks traditionally depends on the building trade.

3.3.1 Analysis of tasks

Ecodesign mainly works in terms of functional sub-units linked to a performance.

Ecodesign's initial focus is to maintain a certain level of performance for the building. In the production phase, it takes a slightly different approach to the setting-up of tasks, by looking at the risks of deterioration.

The object is to supplement the base list with a list of additional control points with the aim of minimizing the causes of failure vis-à-vis the environment.

As underlined during the project design phase, particular care should be made when treating interfaces, which is not especially the case in a trade-based way of working. Individual trades tend to focus on carrying out their tasks and put little emphasis on interaction with tasks done by other trades.

The difficulty thus lies in this much closer coordination between practitioners, which can be facilitated by a functional approach to the work, which organizes it into homogeneous sub-units.

The production transcribes the choices set out in the implementation plans.

For implementation to conform to expectations, each task must be able to draw from graphic documents including all the necessary information.

Box 3.2: Tasks organized by function.

Adaptation	Roads & utility services
	Earthwork
	Foundations
Structure	Floor system
	Load bearers
Envelope	Façade
	Roofing
Partition	Compartmentalization
	Distribution
Equipment	Climatic engineering
	Information engineering
	Movements/mobility
Finishing	Decoration
	Coatings

This work terminates with a package of plans known as the implementation plans. They are different from the design plans because they include the definitive sizing and the definition of all of the components. Structural plans therefore contain not only formwork measurements, but also reinforcement plans and concreting conditions (stop bands, etc.).

These plans are generally supplied by the engineering firm responsible for implementation and commissioned by the construction company. They set out very precisely the production modes specific to the company. Depending on purchasing conditions and worksite practices, choices may differ when it comes to the standardization of installations.

It is also at this stage that the production modes can be envisaged: on-site manufacture, prefabrication abroad or in a factory, etc. Each choice has its own advantages and disadvantages and only the production manager is able to decide, since he is responsible for the financial outcome of his worksite.

3.3.2 Implementation plans

Nevertheless, it is indispensible to supply implementation details, which cannot be improvised or depend on skill alone.

Ecodesign does not change these practices, but supplements them by making it obligatory to supply graphic documents giving details of how items fit together.

For example, airtightness, resulting from a stricter building design that deals with energy losses, will involve making jointing more airtight. This is also the case for individual points that may turn out to be potential linear thermal bridges.

Solutions therefore need to be recommended in the form of drawings illustrating implementation procedures that are integrated into the quality control plan.

Production results from a good balance between the sizing of the constructions and the production capacities.

One of the difficulties inherent to organizing production resides in the supply of the implementation plans. Two scenarios are possible.

The first of these consists in handing over to the technical project management the complete "ready for implementation" file. However, this configuration does not necessarily take account of the production company's specific features. In the second, symmetrical case, the company has the entire responsibility for its implementation plans at the risk of slightly altering the project.

Yet in both situations, it is indispensable to ensure that the ready-for-implementation file is organized with respect for the building time schedule. This pivotal connection between the architectural studies and the start of on-site tasks is essential for the project to be successfully carried out.

The aim of linkpoint planning is to be able to schedule the delivery of plans to take into account the various intermediate approval stages (project manager, control agency, etc.) and the time required to supply and possibly shape the corresponding materials.

3.3.3 Linkpoint planning

Ecodesign entails making choices under multiple constraints.

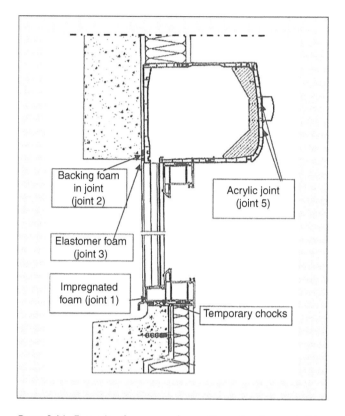

Figure 3.16 Example of measures for making a doorframe airtight.

On this point, ecodesign introduces few changes, except for taking time to reflect on the origin of resources used in the building.

As this stage, it is not the nature of the materials that is at play, but that of the supplier.

Two situations can be envisaged:

– Materials are produced close to the site (local production);
– Supply requires bringing components in from a more remote production site.

In every situation, when the nature of the components makes it possible to do so, a choice will need to be made between the transport distance and the cost of the supplies.

For the same cost and performance, ecodesign favours short supply routes as long as the delivery schedule is compatible with the linkpoint plan. However, transportation wields less weight than may appear likely thanks to continued progress in logistical resources.

No definitive solution therefore exists to organize production, which is a synthesis between diverse parameters.

Worksites are characterized by the care taken in conveying resources.

Making resources available for production involves preparing and organizing vertical and horizontal handling. Moving these items towards the work station is also subject to a plan.

Rough work involves detailed information on the rotations of the formwork panels (forming tools), but other parts of the construction may warrant a study on procuring supplies to ensure that the worksite runs smoothly. In parallel, a study of concrete placing should lead to setting up security bridges.

Regarding so-called light work, preparations are often restricted to the delivery (volume delivered and date of transportation).

3.3.4   On-site transport and delivery of premises

Ecodesign also considers waste produced throughout the production cycle.

In this area, ecodesign puts the emphasis on the by-products generated by the delivery of the premises.

It thus takes into account surplus consumption resulting from production that has not been fully thought through:

– Loss due to handling (breakages),
– Loss resulting from a bad layout plan (unused surpluses or lost cut-offs),

as well as the production of packaging waste (palettes, protection film, etc.).

Unlike in the industrial sector, building work has not yet taken on the practice of fully measuring these items, and often simply accounts for them with a lump-sum percentage applied to the project's bill of materials.

In principle, this information is included in the FDES (French environmental and health declaration) that industrials working on the project must produce for everything to do with resources manufactured outside the worksite. For procedures carried out on site, an effort needs to be made to identify the quantities actually used.

Most of the production involves establishing an environment that makes it easier to carry out all of the elementary tasks.

Once all of the above items have been dealt with, it is possible to carry out the actual realization of the installations, in other words, transform the resources into physical objects. This thus involves productivity and security.

A diagram can be drawn to sum up the production. Seven conditions are indispensible for the processes to run "normally", in other words with no surprises. Although these principles are simple, they are a lot less easy to apply when faced with a multitude of inter-dependent tasks.

3.3.5   Production and security

Ecodesign recommends using "sound" materials.

Ecodesign only brings one additional dimension to this aspect, which is that of health safety.

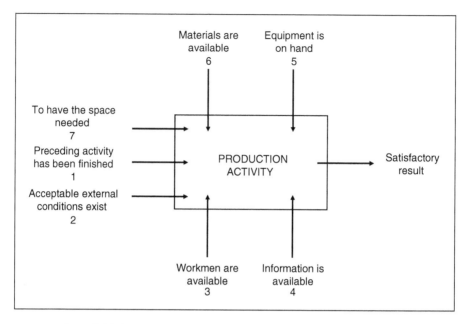

Figure 3.17 The seven conditions required for activities to run smoothly.

Production must mobilize resources that are inoffensive to those using them. This subject can be partially dealt with from a health impact point of view.

However, this comes under the competence of the REACH regulation introduced by the European Commission that applies to construction companies. The regulation's objective is to ensure that materials manipulated on site are harmless.

> *The production of a building is the life cycle phase in which capitalistic intensity is at its highest.*

Even though the building's price is set at the time of the call for tender, it is the worksite balance sheet that enables a decision on the "real" technical cost.

In fact, the production phase corresponds to a partial cycle of the building's life cycle, which is that of investment. The delivery of an operation concludes a whole series of tasks that must be financed by the amount of its assignment to the contractor.

It is particularly important to remember that the appreciation of this process generally results from a succession of operations summed up in the graph below.

3.3.6   Financial aspect

> *Ecodesign moves towards putting a monetary value on the externalities generated by building work.*

Ecodesign indirectly introduces a new chapter into the financial approach.

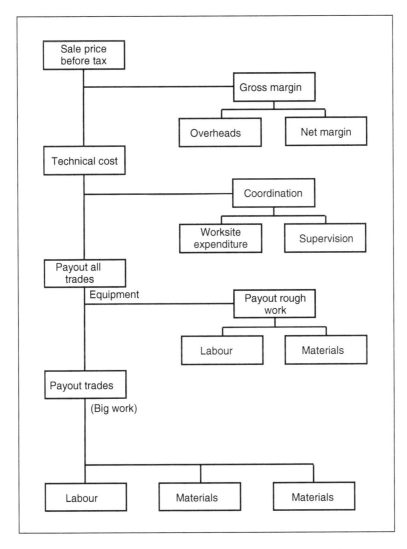

Figure 3.10 Principle of breaking down the sale price of a building.

It authorizes a "carbon" appraisal that is progressively added to each operation balance sheet.

It is not the role of this publication to discuss whether this tax is a good idea, but rather to observe that it is based on a simplified life cycle analysis. Ecodesign is thus a development approach for a new economic model that takes into account the externalities generated by all activities.

The diagram below illustrates the logical interleaving between the conventions for establishing a cost and calculating an LCA.

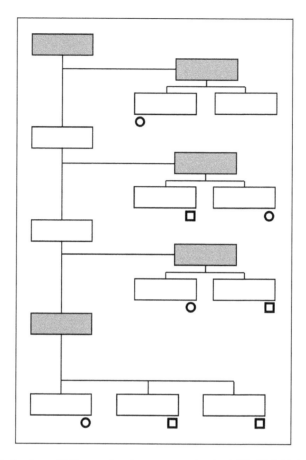

Figure 3.19 Calculation base: □ Flows taken into account by the LCA; ○ Additional flat-rate items due to "carbon" approaches; ▓ Financial items.

3.4   TRANSFER MODALITIES

The transferral of a building to its operators is conditioned by the formidable task of the building's acceptance.

The production phase ends upon the acceptance of the works.

This stage involves verifying whether the construction companies' work conforms to the terms of the contracts signed at the start. It is worth pointing out that the clauses verified here relate to whether the content of the agreements has been respected, but in no case do they concern the building's performance. The only aspect called into question involves the means implemented, which must be identical to the project instructions.

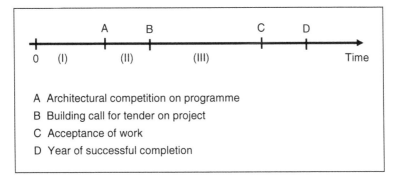

Figure 3.20 Landmarks in the value chain defined by professionals for each of the three phases in an operation.

The conformity with the expectations expressed in the programme is only thus assured in as much as the project corresponds exactly to the programmed terms. In other words, end users must accept the final result without any guarantee directly linked to their expectations.

In this sense, the acceptance is more of a formality between professionals than an approval from the users.

3.4.1 Acceptance

Ecodesign involves the notion of evaluating the result obtained.

With ecodesign, the conclusion of the production phase is envisaged differently.

Given that the works are ruled by performance-related clauses, the acceptance will consist in a series of measures (visual, thermal and acoustic comfort, geometric control and airtightness). The aim here is to ensure that the nominal performance level is reached.

However, experience has shown that it is preferable to wait for one year of operation to verify whether equipment, and in particular consumption of flows (water, energy) are functioning correctly. Additional measures are therefore made one year after acceptance to check the level of effective thermal performance.

These two measuring campaigns taken together are what the experts call "commissioning". The objective of commissioning is to obtain a genuine equivalence between the performances set out in the programme and those resulting from actual usage. This additional operation is undertaken with the sole aim of serving end users.

An environmental approach calls for the production of an additional document, the environmental notice, which has not yet found its base form.

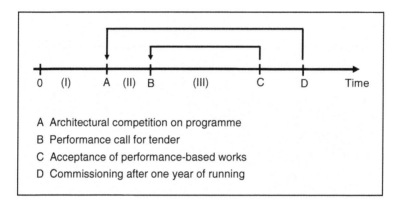

A Architectural competition on programme
B Performance call for tender
C Acceptance of performance-based works
D Commissioning after one year of running

Figure 3.21 Additional verifications included in a performance approach.

Production traditionally ends by handing over an as-built file, which groups all of the data corresponding to the state of the building as observed upon acceptance.

Recently, however, this information has been supplemented with a collection of environmental and health statements (FDES) for all components used in the construction. These statements are designed to make the environmental plan easier to trace and respond to a precautionary concern in the face of any future incidents, such as resulting from asbestos.

The idea is to keep a precise reminder of the resources used in the building to make it easier to target future interventions.

3.4.2 Environmental notice

Ecodesign puts the emphasis on information that can be directly assimilated by future users.

In addition to the environmental notice, ecodesign displays a building's performances.

Ecodesign thus makes it possible to ensure that the initial objectives have been reached and that expectations can be assumed. This communication takes the form of a graphic label providing information on each of the points of the performance class reached.

This method is simply an extension of the practice already in place for household appliances in the European Union.

Given the successive adjustments made by the end of the work, the plans need to be regenerated to reflect the building as it is.

When the operator of takes charge of the building, he needs to possess reliable documents illustrating the work undertaken.

Figure 3.22 The building label system is in the process of being standardized in the EU.

This as-built package of the different implementation plans is based on the definitive state of the operation, and takes into account any changes made to the architect's plans.

This document, which should be put together with care, is essential and constitutes a reference for the continued life cycle of the building, including for any claims that may be made during the period defined in the ten-year guarantee.

3.4.3 Completed project file

Ecodesign is not the object of a specific closing file because its effects have been integrated into the project's execution.

> *The introduction of an advisory maintenance file is devised to underline the importance of maintenance.*

For several years, French contractors have been required to keep an advisory maintenance file up to date (*dossier des interventions ultérieures sur l'ouvrage* (DIUO)).

The main purpose of this document is to set out prerequisites for all work done on the building. The intention is to anticipate any risks facing workers carrying out upkeep or maintenance on the building. It is based on the idea that a better understanding of the way the building is put together will make it easier to determine the best measures to take.

It is worth noting that this requirement is one of a long list of measures drawn up with a view to improving security in building work.

3.4.4 The maintenance file

> *Ecodesign is organized around a building maintenance scenario.*

Ecodesign has no direct influence on maintenance operations except for the fact that it anticipates them.

The emphasis is not so much on integrating a security aspect, but on stipulating the actual nature of the tasks that need carrying out over the long-term.

The only precaution is therefore to memorize these hypotheses so that they can be followed up in the future.

> *Advice on using buildings is only starting to emerge.*

Traditionally, buildings have been considered as "robust" objects that do not require any particular advice on how to use them.

As a result, few operations include instructions for use, with the exception of ventilation systems, which are sometimes accompanied by recommendations for use.

3.4.5 Instruction manual

> *Ecodesign naturally involves using a "scorecard" which is the start of genuine construction management.*

Ecodesign is by definition entirely focused on usage, and thus on the different scenarios used to simulate the way a building operates.

Usage can be broken down into different components:

–   An occupation scenario, describing hours of use and the number of people present on the premises,

-   The temperature settings considered as satisfactory by the users,
-   A ventilation scenario for the premises in volume following a time line,
-   Contributions resulting from the various tools employed by users,
-   The precise use of shading devices over time.

All of these data give the best possible idea of future usage, as long as users conform to these hypotheses.

The instruction manual is thus firstly a scorecard devised for improving usage. It gives a sense of responsibility to both the end users and the contractor, who initially stipulates the users' behaviour and uses it to shape the whole of the operation.

3.4.6   Synopsis

Built objects constitute a support for the activities for which they have been designed. In these conditions, any gap between the objectives put forward/assigned and the way that buildings actually behave will have significant effects on future operating methods. This interaction fully justifies taking great care in each of the different production modes.

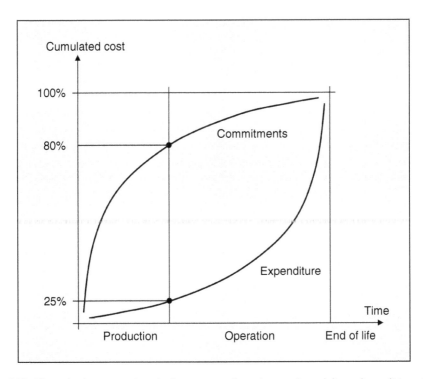

Figure 3.23 Through the care taken in its construction, the product delivered conditions future expenditure throughout its lifespan.

3.5   OVERVIEW GRAPH

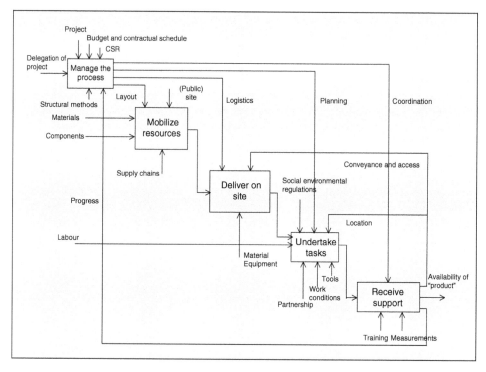

Figure 3.24 The production of a building is a complex process because it is created on top of its own result.

The way that building production is organized is original in terms of the production modes of the industrial sector. Each object produced is unique and is located in a specific context. However, the objective is always to deliver a product whose performances can be verified. Ecodesign fully contributes to this process.

Chapter 4

Ecodesign in the operation phase

The operation phase is naturally entirely centred on the user.

4.1 MOBILIZING THE ACTORS

The end users are naturally central to this phase in the life cycle of a building, for several reasons.

Firstly, end users enjoy the benefits of the built object, which allows them to carry out their various activities. This is the primary purpose of all building projects.

However, the operation phase is also the phase that mobilizes the most resources. Although the investment cost is relatively high, it is nevertheless a lot lower than the operational costs linked to using the building. The graph below illustrates the relative share of the operation for a residential building. This aspect is generally overlooked because the expenditure is spread over time, which lessons its burden.

4.1.1 The significance of end users

Ecodesign must allow usage to be as close as possible to behaviour.

Ecodesign fits in totally with the above-mentioned perspective. It lays even greater emphasis on this phase for end users because it holds them partly responsible for the building's smooth running.

By considering buildings as machines that interact with users, ecodesign thus endows users with a significant responsibility of which they are not always aware. It is, however, undeniable that a good mechanism that is badly used will underperform in comparison with its potential.

This "active" responsibility involves two prerequisites, namely: (1) acculturating users who tend to consider their living environment as a burden rather than using it as a tool, (2) accompanying users to make it easier to manage this support and to render it more efficient.

Buildings require technical assistance over time, like all products.

Whatever the type of building, users can only fully benefit from it with the help of a different group of actors, i.e. the operators.

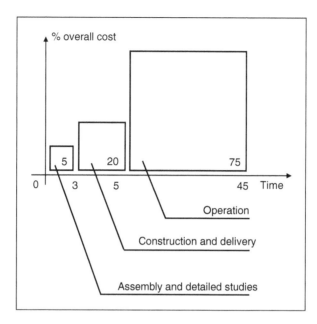

Figure 4.1 The preponderant role of the operation phase in the overall cost.

The term operator comprises two types of practitioner: (1) administrators (property managers, real estate companies, etc.), and (2) suppliers/energy technicians. The combination of the two not only allows the users to "function" as a whole, it also enables each space to be used in acceptable comfort conditions.

All of the interventions grouped under the term operation constitute a spectrum of services that accompany a building's functioning in the long term.

Note that this list gradually expands as needs change, so that it now includes human services (home help, caretaking, etc.).

4.1.2   The operator's role

> *Distributors in the energy domain have a direct relationship with environmental impacts.*

The ambit of ecodesign is restricted to consumption and major upkeep, since these are the main generators of flows during this phase along with the production of waste.

However, even with this limitation, it is important to remember the numerous areas that come under the label of "consumption".

This term illustrates the need to get to know what users are using so as to deduce the most precise volume of needs possible. This exercise is made more accessible with the production of a series of data.

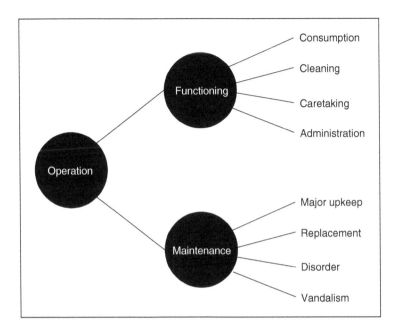

Figure 4.2 Main items of expenditure resulting from operating a building.

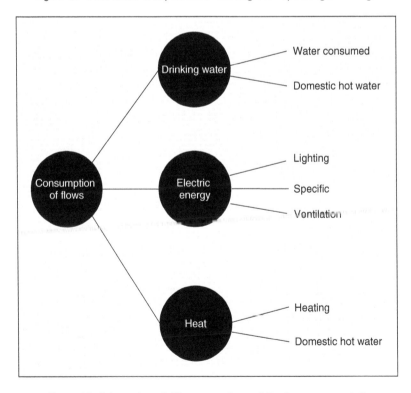

Figure 4.3 Sub-sections fed by connection to "distributor networks".

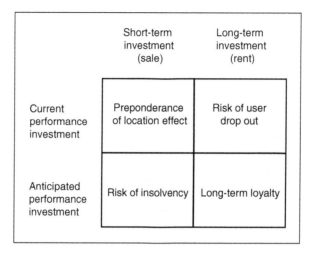

Figure 4.4 Investor strategies influenced by performances.

In the course of its operation, a building necessarily reflects the initial choices made by investors.

Buildings represent significant capital assets for their financers. Two set-ups exist, in particular regarding the operation mode.

Most often, investors are simply looking to place their money. Sometimes, however, investors operate buildings themselves. In this case, their concerns will be as set out in the previous section.

In the first situation, several investment strategies are possible, as illustrated by the following table. This table clearly indicates how the decision's time horizon conditions the level of the performances assigned to the operation.

This choice is increasingly important in relation to sustainable development requirements, which are having the effect of profoundly changing market conditions (risk of taxation).

4.1.3 The investor's role

Investors can take advantage of an ecodesign approach to guarantee their choices.

Ecodesign has a particular status vis-à-vis investors, which is that of a new resource. This "measuring" approach gives investors an opportunity to:

–   Communicate better about the nature of their investment by announcing a supported, significant environmental profile;
–   Establish a genuine benchmark to ascertain the real value of their investment vis-à-vis other similar operations (whereby performance is a gauge of efficiency).

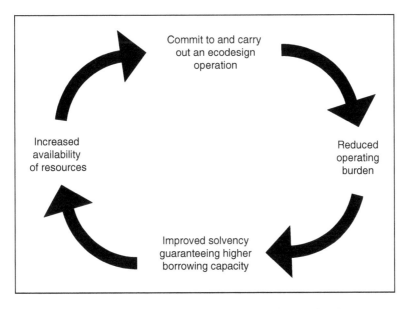

Figure 4.5 Virtuous circle of ecodesign on a financial level.

In addition, ecodesign allows a certain capacity for steering, or more precisely evaluating, user behaviour and thus being in a position to enter into more constructive dialogue.

> *The operation of a building generates tasks that in turn create economic activity.*

In the building sector, the issue of finance is crucial. This is due to the long-lasting nature of the product and the considerable amount of money involved, even without counting the land.

A complex organization is put in place, involving the banking sector, insurance and investment funds.

This complex system is most active in the operation phase.

4.1.4 The financing burden

> *As a principle, ecodesign sets out to be the driving force of an improvement process that must generate a financing capacity.*

This aspect of a building operation should logically be influenced by the use of ecodesign.

The diagram below provides a clear illustration of this. In ecodesign, a virtuous cycle is likely to occur, unless it is overshadowed by the more powerful phenomenon of land income.

Numerous studies have been carried out in Anglo Saxon countries on the added value resulting from higher environmental performance. The conclusions are not decisive, although the indication is that a general trend is starting to take place.

In particular, ecodesign could be established as one of the criteria that encourages responsible investment in the banking sector.

A building's operation is conventionally reduced to regulation by the community.

Since the operation of buildings involves human beings, it attracts attention from civil society representatives. The result is fairly significant regulation of the act of building.

However, it is worth stipulating the nature of this action as it stands today.

Two schools of thought exist. The first, which is favoured in France, consists in checking whether processes conform to archetypes that are supposed to render best practices. The second, more commonly used in English-speaking countries, focuses on measuring significant performances, in other words, those perceived as such by users.

Each of these two methods has advantages and disadvantages. It is worthwhile considering them in parallel.

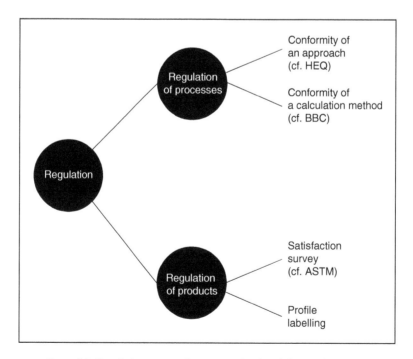

Figure 4.6 Regulation covers the process level and the product level.

4.1.5 The incidence of regulation

Ecodesign can result in building "labels" that implicitly introduce a dialogue with users.

The issue is also raised by ecodesign, which can be used in two ways.

As a thought-out approach, ecodesign makes it possible to justify the choice of a solution. It allows the gains obtained to be visualized with a variant of a common construction mode. This explanation illustrates a capacity for criticism and a desire to achieve genuine optimization from an environmental point of view.

With its capacity for simulating how the building will operate, ecodesign encourages aiming for realistic performances. This precaution in terms lies in the fact that a modelling of the "construction" system can be carried out taking users' behaviour into account. However, the more precise the usage scenario, the more closely the calculated performance will converge with the performance experienced.

Clearly, given the significance of the operation phase, users are primarily concerned with actual performance, which is immediately translated into a burden whose weight will depend on the effective level of the building's behaviour in response to appeals from the users.

An operation's remuneration is highly dependent on its contractual framework.

An operation has multiple contractual forms that depend on stakeholder scenarios.

Currently, two forms predominate: case No. 1 whereby users buy buildings for their own use, and case No. 2, in which users rent from a lessor. In both configurations, usage also entails the purchase and supply of different types of energy from distributors.

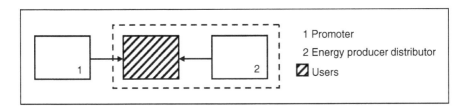

Figure 4.7 Scenario 1: Building purchase for own use.

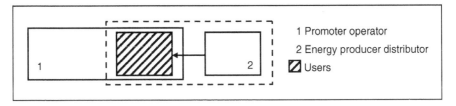

Figure 4.8 Scenario 2: Renting for temporary use.

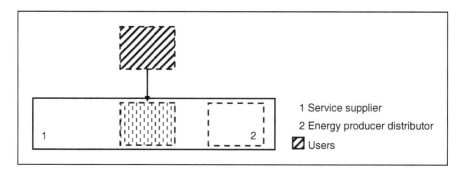

Figure 4.9   Scenario 3: Purchase of a service.

In the long term, it could be possible to envisage an even more integrated scenario in which users pay for an all-inclusive service (functional efficiency).

4.1.6   Contract configurations

Ecodesign can be applied in all configurations but works best in certain types of relationship between users and building suppliers.

From this point of view, ecodesign also brings in factors that reinforce each of the transaction types mentioned above.

In the case of rent, ecodesign sets out the bases of a credible appraisal of future expenses in the form of a clear set of instructions. However, the decision of performance level is directly linked to the investment capacity, given that the profit on expenses does not enter into the profitability calculation base.

The ideal configuration, which is currently rare, is that of a "concession". This is the only way to strike a genuine balance between the investment and the operation. Some specialists prefer to qualify this situation as functional efficiency, since services are not delegated, as the term concession literally implies, but rather services are supplied with optimum efficiency by the supplier.

4.2   OPERATION PHASE OPTIMISATION

Traditionally, users are entirely responsible for monitoring buildings.

The main objective of this phase of the life cycle is to allow the building to be used by the users.

However, traditionally, this is a fairly conventional concern. The ultimate aim is in fact the rate of occupation rather than quality of use. Put slightly differently, we might say that a different final usage must be content with what it has access to, in other words a physical place that is essentially characterized by its geometry and location.

In fact, buildings are all too often reduced to the status of a static object without any concern for the human/machine interface.

Box 4.1: Organization of a scorecard for a building in operation.

1. Nominal performance monitoring

1.1 Comfort temperature:
    Measure of actual inside temperatures
1.2 Humidity rate:
    Measure of internal rate
1.3 Number of hours of discomfort:
    Hours when the temperature exceeds set temperature
1.4 Volume of VOCs:
    Verification of concentration rate

2. Operational performance monitoring

2.1 Energy consumption:
    Level of volumes actually consumed
2.2 Water consumption:
    Level of volumes actually consumed
2.3 CO_2 contribution:
    Evaluation following entries 21 & 22
2.4 Volume of waste produced:
    Correspondence with service company measurements

4.2.1 Scorecard

Ecodesign introduces rigorous monitoring using a scorecard.

Ecodesign, in focusing all of its efforts on evaluating performance, on the contrary integrates the fact that the properties of buildings alter over time.

For this reason, ecodesign includes the principle of a scorecard whose aim is not so much to control operations as to be aware of any changes that might need correction. The building's behaviour is thus monitored more thoroughly.

Currently, eight monitoring values are used which must be close to the performance objectives set out in the operation's initial programme. The table below presents the two classes of monitoring, based on the functional unit.

Buildings generate consumption invoiced to the users.

However, throughout its lifespan, a building is similar to a machine that consumes a certain volume of resources for which users must pay.

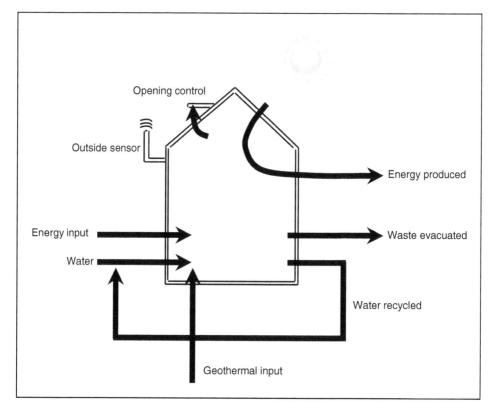

Figure 4.10 Points for metering or collecting information on a building's operation.

This practice usually takes the form of invoices for energy (electricity, gas), water, and in some cases the volume of waste produced. All of these quantities are metered and read at different times.

Yet other flows could also be measured and certainly will be in the near future (c.f. application of carbon tax). However, in all cases, the object of measurement is centred on recording a physical quantity rather than on whether the service is satisfactory. The focus here is on an economic transaction in the form of invoicing for supplies.

4.2.2   Instrumentation

Ecodesign should be accompanied by widespread metering, requiring new instrumentation.

Ecodesign, in setting out performance levels, introduces a need for instrumentation that can be broken down into: (1) a capacity for acquiring data for monitoring and, as a result, (2) a capacity for managing equipment.

Box 4.2: List of data that require measurement.

Behaviour	– Presence sensor
	– Entry sensor
Climate parameters	– Recording of outside temperature
	– Recording of inside temperature

Inflows	– Water metering
	– Electricity metering
	– Gas or heat metering
Outflows	– Volume of hot water
	– Volume of waste sorting
	– Volume of energy sold

For the first aspect, meter technology has made great progress so that meters can be installed non-intensively at a reasonable cost. They can be backed (monitoring value) by radio.

The second aspect includes software packages that can be configured in very varied ways and make targeted use of the data collected. They are interfaceable with regulation tools to make optimal use of the equipment in a building.

Note that this instrumentation is applicable to both new buildings and existing buildings, which can be made more efficient as a result.

Usage performances are still very rarely verified in situ.

The operation phase can only begin after acceptance of the building, in other words after the building has been verified as conforming to the specifications.

However, this procedure is mostly defined by its contractual aspect. It does not necessarily entail a verification of the entire product in operation. A good illustration is provided by comparing the prevailing practices in the naval industry.

In the naval industry, the acceptance process takes place in two phases. The first of these involves an audit of all of the sub-units. It is an acceptance test that entails verifying each component. The second phase consists of large-scale tests that prove that the structure can be manoeuvred at sea. This entry into active service is known as commissioning.

This kind of practice is beginning to be used for buildings when starting up equipment and heating-ventilation-air conditioning.

4.2.3 Commissioning

Ecodesign introduces regular evaluation of the performances initially indicated.

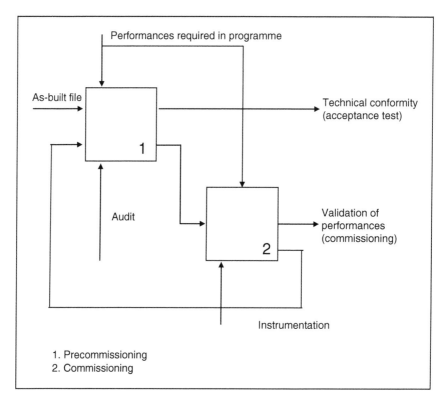

Figure 4.11 Organization of commissioning.

Ecodesign, which views buildings as whole systems, requires that performances announced at the programming stage are effective at the start of the operation phase.

It is based on the widespread practice of commissioning for the whole building. The aim is to validate that the actual operation corresponds to the expectations of future users. However, this approach is a lot more complicated in buildings due to variations in climate and an indispensable run-in period for the different components (e.g. concrete curing).

Building commissioning therefore takes place at least one year after delivery and consists in measuring a number of behaviour values that must be treated so as to isolate the intrinsic performances of a single building. The following diagram illustrates the principle.

The list of measurement points is the same as above.

> *Buildings are supposed to be long-lasting by nature and able to adapt to all practices.*

In standard processes, users' involvement is at a minimum. At the most, it enters into three technical contributions.

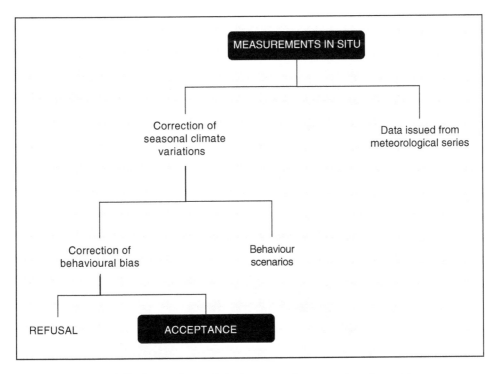

Figure 4.12 Stages of commissioning: successive processing of raw data.

The first of these is supplying a maintenance plan indicating the main tasks to accomplish and their implementation date. However, this document is only effective if it is actually memorized.

The second consists in a certain capacity for managing the atmosphere thanks to a management installation (building automation systems). However, the user friendliness of these tools still needs work, and their problematic record holds back their development.

Buildings are usually only accompanied by an instructions booklet that is supposed to explain how certain equipment works (in particular mechanical ventilation). Nevertheless, this information is generally only partial and its literal character limits its scope.

4.2.4 Accompanying users

Ecodesign mobilizes users to help them make efficient use of buildings.

In ecodesign, users play an important role. This is not because ecodesigned buildings are intrinsically "efficient", but because the use made of the buildings is the result of interaction.

Table 4.1   Extract from a casebook on the lifespan
of a building's components.

	DV (years)
Roofing:	
– Multilayer waterproofing	14
– Asphalt waterproofing	30
– Slate tiles	50
Interior coatings:	
– Paint	8
– Carpet tiles	10
– Thermoplastic tiles	20
– Tiling	50
Heating – air processing:	
– Regulation	8
– Extractor	10
– Condensation boiler	14

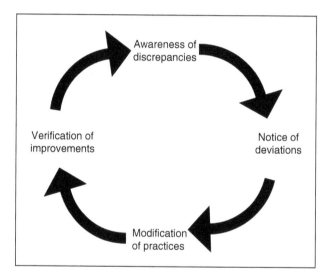

Figure 4.13 The "virtuous" circle of learning and participative commitment from users.

In order for this relationship to be satisfactory, three accompanying measures should be put in place:

–   Users should be made aware of how to make rational use of the building. This mainly involves producing more user-friendly instructions, i.e. less technical and more visual.
–   An early alarm system should be set up for deviations. Prevention involves recommendations illustrating discrepancies and how to reduce them.
–   A detailed evaluation of the cost of each practice should be produced including strict penalties for incorrect behaviour.

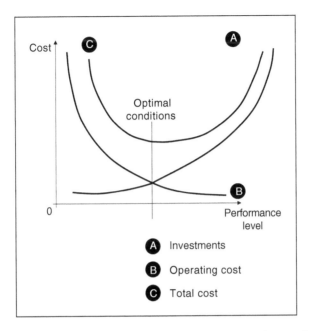

Figure 4.14 The standard issue of maintenance costs to reach an optimum.

The objective of these measures is to sustain the gains obtained and avoid what experts call the "rebound effect", i.e. a drop in vigilance and a return to lax usage.

The operation usually results in a total cost approach.

The operation phase necessarily implies secure use and a capacity to ensure performance without harming the environment.

However, it is indispensible to take this definition further by reflecting on the way in which the result is achieved. The traditional strategy can be summed up with an illustration of so-called total cost optimization.

This approach is based on the following principle: the higher the agreed investment is, the lower the maintenance costs will be. The idea is that sustainability necessarily involves using more elaborate resources

In common language, this approach is summed up by the expression "you get what you pay for". However, this is only true when techniques are in a steady state. In other words, this analysis does not take account of obsolescence.

4.2.5 Challenges

Ecodesign significantly extends the standard finance calculation perimeter by bringing in different environmental impacts.

Ecodesign is based on a much more realistic attitude that does not exclude the fact that buildings necessarily require preventative maintenance due to natural wear resulting from use.

The challenge therefore encompasses two aspects:

- A conviction that the reasonable use of a building that has been carefully devised and constructed is the best option for our living environment and the cost burden.
- A concern for maintaining this balance throughout a building's lifespan, which involves upgrading at regular intervals.

Striking a balance between user behaviour and the performance levels of a building is not easy, partly because no pre-established solutions exist, and partly because it depends on the responsibility of the different users (contractors, operators, users).

The traditional notion of total cost (investment + costs) therefore appears very limited in the face of a much more subtle mechanism that associates dynamics that change over time.

Ecodesign lends fairly significant weight to the readiness of each stakeholder in an operation. This cannot substitute individual commitment, but it presents greater potential for progress than traditional practices.

> *The operation, depending on its configuration, can lead to contrasting views of the financial dimension.*

The operation phase is highly influenced by the type of contract between stakeholders. Depending on the organization between them, the result will be different in terms of the holding cost.

The standard contractual organization in the case of a collective residence is as shown below in configuration 1. This corresponds to a clear dissociation between the right of use and the operating costs. Note that users are not entirely free to choose the distributors, which will have been selected by the investor or his contract manager when the building was constructed.

In this case, an administrator manages the routine maintenance and the contribution due from each user. The use made of the building is thus totally dissociated from its intrinsic performances. As a result, optimizing the holding cost is difficult if not impossible.

4.2.6 Forms of valuation

> *The operation optimization that results from ecodesign can lead to new economic models in the long term.*

Ecodesign does nothing to modify an operation's contractual configuration. On the other hand, it may induce the use of other methods.

Ecodesign puts particular emphasis on optimal operation through close joint work between the builder and the user. This state corresponds to the conditions for genuine optimization. But the holding cost is then considered as equally shared between the contractor and the user.

To arrive at this result, the only possible set-up is configuration 2 shown in the graph on the previous page. This involves a single contract between the operator (who provides a place with the associated comfort conditions) and the user (who agrees to use this support within the bounds of responsible operations).

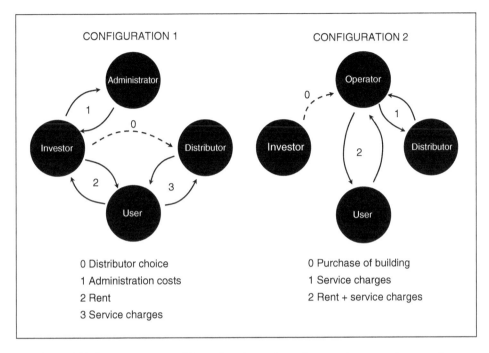

Figure 4.15 It is important to differentiate the notions of rent and payment for a service.

No comprehensive scientific study exists as yet to determine the potential gain of such a configuration. However, analyses of the "rebound" effect (i.e. drop in vigilant behaviour) indicate that a minimum threshold of 10% of energy costs is realistic. Combined with a reengineering approach (i.e. reduction of investment cost for higher performances), the holding cost could logically drop by an equivalent amount.

4.3 REHABILITATION

Rehabilitating buildings is inherent to the ageing phenomenon.

At some point in the operation phase, the contractor will necessarily deliberate on whether to upgrade his building or part with it. There are several reasons for this.

Most often, outside factors will modify the choice criteria. For example, a rise in the cost of energy can lead to reassessing the thermal performance of a building that consumes too much. Another reason is that developments in techniques may render a building obsolete. Information network connections provide an interesting example. Cabling has simplified power supply connections, and WIFI is tending to override them. Buildings are thus ageing prematurely due to technological progress.

However, buildings also age as a matter of course. This ageing depends on numerous combined factors that depend on the conditions of use (careful use or not) and upkeep (preventative or curative). In particular, different components do not have the same lifespans. It is accepted that indoor paintwork will be renovated more quickly than outside woodwork.

The decision to preserve and rehabilitate a building thus depends on several parameters that are linked to the building's attractiveness (location, relative performances compared to other buildings), the financial balance attained, and the economic situation at the time.

4.3.1 Decision factors

> *In ecodesign, a rehabilitation scenario is defined right from the construction phase.*

Ecodesign is concerned by such a decision from two points of view. One concerns upgrading the building over its life cycle. The second is linked to launching a new study.

In the first case, the aim is to ensure that the moment the decision is made corresponds to the scenario selected to carry out the LCA. The work envisaged should also conform to the hypothesis that was agreed on. In general, maintenance work, even when heavy (e.g. changing outside woodwork) is compatible with simulations because it corresponds to a reasonable policy of upgrading to the initial standard.

A more radical transformation seeking to modify the level of nominal performances thus involves launching a new ecodesign approach for the rehabilitation. This second case involves an overall upgrade corresponding to a programme change.

> *Rehabilitation works on an existing building whose state must indispensably be evaluated.*

To undertake a rehabilitation, it is indispensable to carry out a diagnosis. This can be more or less far-reaching depending on whether the aim is to improve a single performance or focus on several performances (e.g. thermal, acoustic, accessibility, security, strength, etc.).

The aim of this diagnosis is to determine the current state as precisely as possible to provide a reference point that can later be used to establish what work needs to be done to improve the level of the performance in question.

This task involves a degree of expertise since the ageing of a building is not always visible and surveys may be required to ascertain its exact state.

4.3.2 Carrying out a diagnosis

> *The components of a diagnosis are used to establish an environmental profile of the existing state that will be used as a reference to measure future improvements.*

To undertake an ecodesign process, a diagnosis is crucial. The data collected can be used to make an ex post simulation, in other words to set the calculation tool with the information collected.

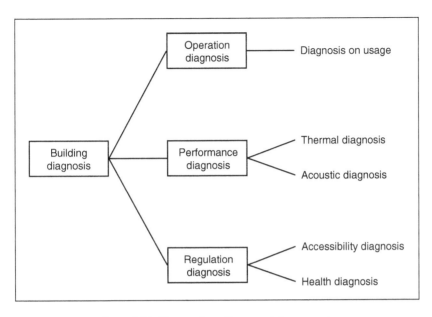

Figure 4.16 How various diagnoses interconnect.

For energy in particular, it can be useful to compare dynamic thermal simulations with the actual measured consumption. This comparison is the most reliable way of ascertaining the real performances of the existing envelope (especially air infiltrations).

> *Extensive rehabilitation constitutes a project in itself that follows the same precepts as a new building.*

Over the years, it has been more or less established that to be financially viable, the energy-efficient upgrading of existing buildings requires the broadest possible base. In these conditions, the actual functional features of the building are at play.

The objective in this case is to redefine the use of the location. In some ways, it is a kind of urban reconstruction. This involves launching a comprehensive project along the same lines as the recommendations put forward above. More precisely, a detailed programme needs to be drawn up.

4.3.3   Changes in functional features

> *Ecodesign makes it possible to optimize rehabilitation by validating technical and architectural choices.*

Ecodesigning new buildings involves comparing projects that respond to the same operating specifications.

In rehabilitation, the context is determined in advance with an existing site inventory. An ecodesign approach involves deciding which of the different refurbishment hypotheses is the most environmentally favourable.

To sum up, the functional unit is defined by the limits of the existing structure and the planned adjustments. Ecodesign involves setting out the differences that exist between the possible technical and architectural solutions.

> *The list of rehabilitation techniques is very broad and almost as varied as for new buildings.*

The technical solutions to be implemented fall into two main categories: one concerns the interior layout and the other corresponds to improving thermal performance.

For the first category, the technologies used are the same as for new buildings. However, those relating to thermal performance require more care, in particular when a high level of performance is sought.

In this case, the corresponding cost is considerable and the objective is to work out the best performance/cost compromise in conjunction with an investment policy. The issue is thus to invest, but not more than a certain amount determined by the residual value, while paying attention to the future, i.e., without ruling out the possibility of a future rehabilitation (step-by-step rehabilitation method).

4.3.4   Technical solutions

> *An ecodesigned rehabilitation should be thought out in terms of a progressive scenario of the action to be done. The objective is to never exhaust the remaining potential by making choices that could jeopardize it.*

Ecodesign is particularly concerned with saving energy. It focuses more specifically on optimizing solutions. These choices, without altering the usage scenario, will involve highly contrasting performance levels.

The following table shows the progression of the action to undertake without undermining the remaining potential.

> *Rehabilitation consists in retrieving or attaining a performance level whose effectiveness should be verified.*

By definition, rehabilitation involves restoring or upgrading a building. Nevertheless, the precision of this redevelopment raises a new problem, which is how to measure the expected results.

The issue here is the gap between the existing and planned states. It is two-fold since it concerns two states.

The initial state is subject to a diagnosis as indicated above. However, this step is not sufficient. Based on the data collected, it is indispensible to simulate the building's

Table 4.2 List of possible action classed in rising order of performance when cumulated.

1. Act on regulation	– Take intermittence into account – Refine regulation – Lag distribution channels – Use energy-saving lamps
2. Act on the envelope	– External insulation of opaque parts – Insulation of basements – Insulation of roofing – Optimization of glazing – Architectural elements (verandas, oriels, etc.)
3. Act on equipment	– Change heat production sources – Recuperate heat from ventilated air – Renewable energy gains – Manage lighting (presence sensor)

actual level of performance. Technical improvement solutions can then be calibrated in line with this result.

The aim of a responsible approach would be to carry out a comparison (relative gain) using a calculation tool that is as close to reality as possible. However, for a range of complementary reasons (subsidies, loans, taxes, etc.), this process should be combined with the display of an absolute, so-called regulatory value.

The problem that arises is thus to ascertain whether two tools need to be used, or whether one tool could be sufficient, given that, when it comes to energy at least, the operating performance is translated by a cost that the contractor wants to know as precisely as possible.

4.3.5 Measuring method

Ecodesign is organized around software that can be used to closely anticipate future results.

Ecodesign, in particular for rehabilitation, only brings added value when it makes it possible to get closer to a building's actual behaviour.

One question then arises, which is the means used to calculate energy performance. There is no real benefit in making do with a simple regulatory calculation. The objective is to anticipate actual consumption levels as precisely as possible.

Only one approach is realistic in this case, which is a dynamic thermal simulation based on a usage scenario thought out with the contractor and based on the most plausible hypotheses given the financing envisaged.

Ecodesign in rehabilitation thus becomes a crucial management method given the emphasis laid on the energy aspect.

Traditionally, a rehabilitation operation begins with an increase in the building's asset value.

The purpose of rehabilitation is to obtain a building whose expected resale price is higher than it was in its initial state.

In this sense, the acquired value can be qualified as an investment since for the contractor it represents capitalization, a kind of assets saving.

The accent is thus on the built object envisaged as a financial vehicle.

This approach is part of a short-term perspective and feeds into a real estate market that is all too often restricted to its speculative dimension.

4.3.6 The residual value issue

The objective of an ecodesigned rehabilitation is to increase its residual value, or in other words improve its performance level.

Apart from the above-mentioned aspect, ecodesign puts the accent on improving the operating mode. The objective is to make a space available to users that is more respectful to the environment for a considerable length of time.

This usage value, which can be defined as a residual value, is not yet fully recognized given the importance of the assets. However, enlightened investors seeking to develop a genuine benchmark among the various real estate offers are starting to take it into consideration.

In the long term, this trend is likely to become commonplace in the building sector, thus aligning construction with consumer goods. It would also constitute a new framework for managing the industry.

Rather than a fallacious utopia, this is an inevitable extension of usage value criteria, taking environmental issues into account.

4.4   EXPERIENCE FEEDBACK

At present the building industry does not take advantage of capitalizing on feedback.

Buildings are not exempt from the rules of good industrial management. The same can be said for taking advantage of acquired experience, which specialists call feedback.

However, this approach is developed within a fairly limited perimeter, which is the lifespan of the building's various components.

There are several reasons for this, such as continuity within the industry, diverse contexts (location, uses, etc.) and the duration of the life cycle, which is very long and makes continuous genuine traceability more difficult.

4.4.1   Recording traceability

Ecodesign involves making better use of feedback.

In ecodesign, collecting feedback is both easier and more wide-ranging.

Since ecodesign involves instrumentation, storing the results of measure samples is more natural. The drawback is that the data is collected but not processed: the

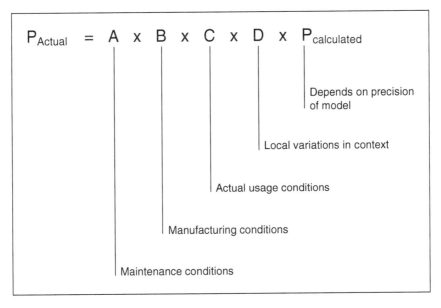

$P_{Actual} = A \times B \times C \times D \times P_{calculated}$

Depends on precision of model

Local variations in context

Actual usage conditions

Manufacturing conditions

Maintenance conditions

Figure 4.17 So-called factor method taking into account the different parameters of a performance's deterioration.

most useful information is in the analysis made between the variations of different parameters.

As pointed out several times above, the measurement field is very important and can be defined as genuinely multidimensional. It is this wealth that needs to be exploited. Some practitioners are starting to undertake this work, although for the moment it is still rare.

In Europe, some countries have organized the obligatory collection of data relating to buildings' behaviour with the aim of introducing a benchmark and in particular verifying that the labels awarded correspond to genuine progress in the operating phase.

> *Components used in rehabilitation should all be the object of an identification sheet in the long term.*

Generally, buildings are essentially viewed as the sum of a very large number of elements. However, this attitude is not neutral.

In particular, environmental requirements have been translated by a fairly mechanical response, which is to simply supply data per element. This task tends to be outsourced to suppliers, with the result that industrials in the building sector have established environmental product declarations (FDES in France). The popular notion is that the availability of these declarations is enough to ensure the environmental performance of the result, which is thus no more than a verification.

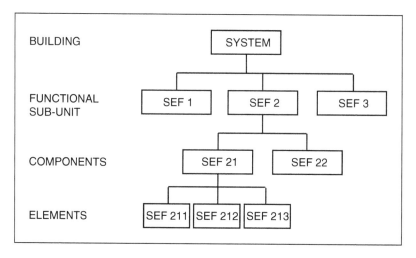

Figure 4.18 Construction broken down into single elements.

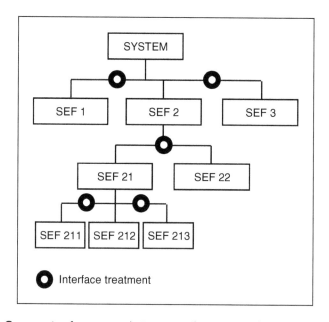

Figure 4.19 Construction from an ecodesign point of view: a set of interconnecting elements.

4.4.2 Availability of FDES

However, ecodesign involves a more precise analysis of the junction between structural elements.

In ecodesign, the approach is more systemic. It consists in looking at what happens to the interfaces during a project's life cycle. It is thus necessary to make a much more detailed examination of all of the progressive organization levels of a building.

In principle, the question that arises is how to devise LCAs at each stage. The objective is to harmonize the calculation protocols at each level (perimeter and scenarios). However, this involves significant rigour and transparency, which is only achievable through common discipline standards accepted by all.

The keystone of this process is a sound knowledge of the interface treatment, in other words, a capacity for making a crossed analysis of building trades.

> *A building can no longer be considered simply as a static physical object.*

In a no doubt rather simplistic manner, traditional buildings are sized to be fairly robust in the face of often very variable cycles of use.

Put another way, the method that has predominated for around half a century is to seek a fairly resistant built environment to compensate for users' hieratic behaviour.

In the light of the new requirements of sustainable development, this policy is probably not the most economical, since it necessarily results in oversized buildings.

The whole issue is therefore to cover this surplus cost, which could be termed the cost of solidity.

4.4.3 Scenarios

> *The major advantage of an ecodesign approach is the optimization attained within a usage scenario accepted by all stakeholders. It is crucial to adjust these scenarios to match real-life experience and enrich them with feedback.*

In ecodesign, a central assumption is rooted in the practice of value analysis.

This industrial design method stipulates that a bare minimum exists that corresponds to user expectations set out in detail and fully assumed over a pre-determined time period.

All tools used in ecodesign are based on this same principle because they result from the same functional unit and reflect a precise usage scenario. This is the case for thermal simulation tools and LCA calculations.

This "bare minimum" results from an inter-disciplinary work that brings together all stakeholders in an operation. It is not absolute and directly depends on the expertise of practitioners involved. In other words, the result sought will vary depending on the context. However, it will not be any less efficient, as illustrated by the work done on concurrent engineering.

Nevertheless, over the entire life cycle, the total result is highly dependent on the users. If users want to attain the results announced, they must conform to the selected scenarios.

It is clear that choosing this approach depends on individuals' readiness to take on responsibility. Acceptance is becoming easier with the awareness of environmental issues and lower risk of general indifference on the subject.

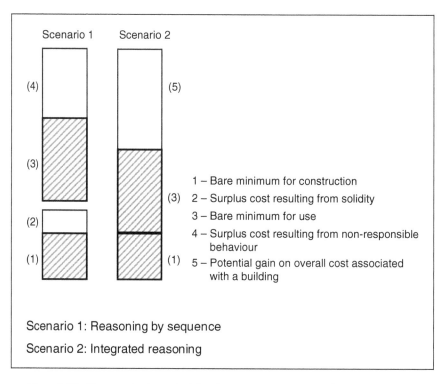

Scenario 1: Reasoning by sequence

Scenario 2: Integrated reasoning

Figure 4.20 Characterization principle of two scenarios to consider a construction.

> *If a comprehensive set of FDES were established covering all of the construction products, it would constitute a coherent whole. Practice shows that this is not yet the case and that means are required to rectify the situation.*

From an environmental point of view, numerous professionals consider that FDES for the different components of a building suffice.

The issue is not however so simple, because a building cannot be reduced to the sum of its components. In addition, the set of environmental declarations does not yet cover all solutions implemented, and only concerns the manufacturing phase.

What is more, each declaration also corresponds to an LCA produced by the supplier. Few people are currently qualified to analyze the protocols for putting together these documents, which often require databases that are not necessarily French.

What emerges as indispensible is achieving genuine traceability of the operating mode so as to be able to work in a transparent manner.

4.4.4 The impact matrix

> *A matrix made up of the impact factors of the different components of a building will never be definitive because it corresponds to a given*

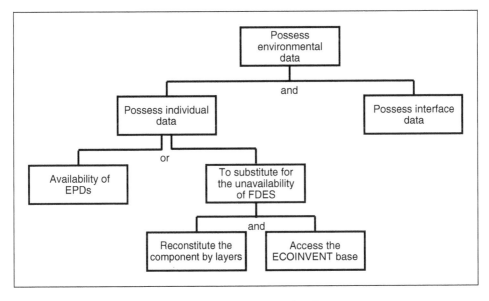

Figure 4.21 Data acquisition logic for all projects.

state of scientific knowledge. It is therefore essential that it should be displayed from the start of any ecodesign project.

Ecodesign is based on the impact matrix. This table, which is the consolidation of scientific expertise, directly depends on the level of knowledge shared by specialists.

Historically, the first attempts were made by ECOINVENT, which advocated comprehensive treatment of the components of a building rather than exhaustive coverage of individual items. About ten years later, French building industrials established the FDES protocol. This work is underway and its coverage rate is gradually extending.

The question that arises is how to combine these sources of information in order to successfully establish LCAs of buildings, and do so explicitly while awaiting a single, comprehensive inventory for the country.

This is the only procedure that allows for progress, given that the resulting gaps are real, but reasonable in size, as shown by studies carried out at European Union level (PRESCO, REGENER).

All buildings are based on an evaluation in line with standard practices between professionals.

In practice, buildings usually result from an adjustment that aims to strike a compromise between performance and budget. This compromise is subject to a life cycle marked out by calculations that are regulatory in as much as they define the minimum conditions for durability valued by the community.

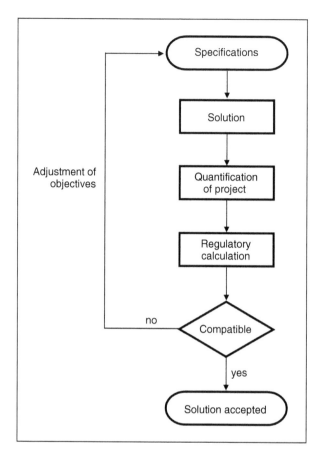

Figure 4.22 The traditional way of thinking when planning a building.

This standard economic model (i.e. a business-as-usual scenario) is perfectly justified when applied to traditional criteria, i.e. ensuring the conditions for sustainability. It is more debateable in the hypothesis of extending the criteria to all environmental impacts.

4.4.5   Evaluation tools

From an ecodesign point of view, the objective is not just to obtain a reliable result, but also to justify that the choice made is the best option in the circumstances.

Ecodesign makes use of simulation tools that attempt to apprehend a building's operating conditions as realistically as possible, provided that the soundness of the project has been established based on traditional criteria.

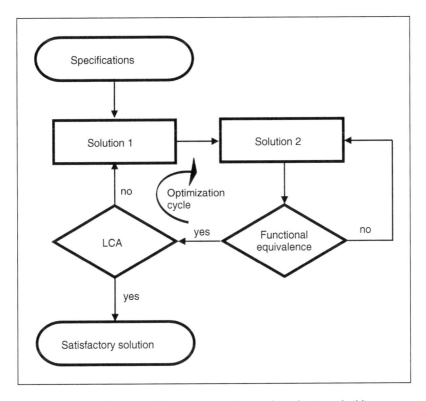

Figure 4.23 The way of thinking in ecodesign when planning a building.

The second characteristic of ecodesign tools is that they can be used for comparative studies. The objective is to propose ex ante a solution that has been checked either against an existing reference solution or a possible variant. The aim is to ensure that the choice made is the best one, all else being equal.

The direct consequence of this attitude is that this choice-making method becomes more refined over time provided that experience is capitalized and corresponding data can be reliably traced.

4.4.6 Synopsis

The object of ecodesign is to help the operation phase run in line with the hypotheses that presided over the building production.

Some experts qualify this result as the quest for a high eco-efficiency level, i.e. an operation in which the environmental performances are improved, all things being equal elsewhere.

It involves instrumenting the building to ensure that measurements are part of the construction management.

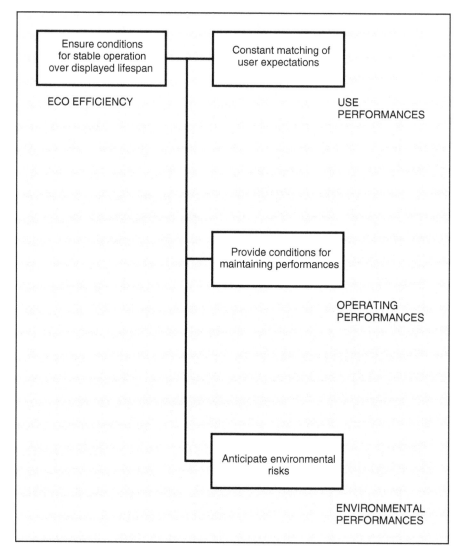

Figure 4.24 During the operating phase, ecodesigned buildings are less subject to the hazards of a "standard" building because they have anticipated behaviour. This reduction is a gauge of reliability and thus savings.

4.5 OVERVIEW DIAGRAM

Its implementation during this period can be broken down into two main parts:

- Collection of data to be used on other projects.
- Rehabilitation, which mobilizes new resources that also need to be chosen in terms of their environmental impact.

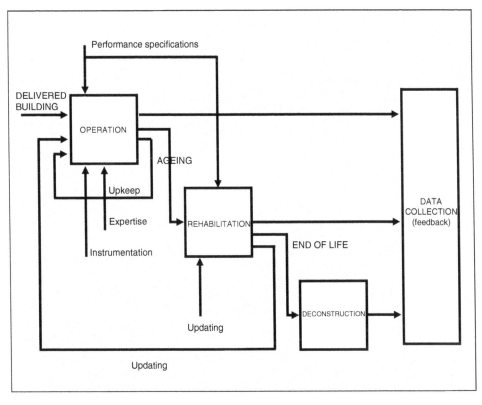

Figure 4.25 This result is the consequence of a genuine capacity to simulate a building's life cycle, i.e. an improved reasoning of how it will be used. This can only be obtained when monitoring can be tracked and data capitalized.

Yet it only really comes into its own when it makes use of feedback on experience.

Conclusion

In fine, how can we characterize ecodesign?
What are its capacities?
What developments are in view?

The information collected on ecodesign can be summed up by answering three questions that together sketch out a conclusion.

The first question relates to qualifying this recent approach to building. Should it be considered as an additional methodology come to join the ranks of standard practices? Or, on the contrary, is it a reworking of the way of thinking? A more fruitful middle way would no doubt be to use ecodesign as a way of enriching and opening up the scope of possibilities in the art of building.

The second question relates to the impacts that this kind of approach could have on technologies. Could new technical procedures emerge? Is it possible to envisage a specific family of ecotechnologies? Put a different way, it is indispensible to be able to describe the specific capacities of ecodesign.

However, ecodesign is still being developed in the building sector. It is now necessary to draw up an inventory of the next steps for spreading the practice among professionals. The approach of viewing our living environment as a system opens up a very rich scope of possibilities, but exploiting this potential will call for new efforts both on a theoretical and practical level.

In conclusion, from this last analysis, ecodesign emerges as the expression of a new impetus for the building industry that can provide a response to Auguste Perret's vision: "A building is a means by which architecture should take shape".

Ecodesign is primarily a way of optimizing building projects.
Ecodesign can be used to explicate the choices presiding over a project.

The emergence of environmental issues has an incidence on the behaviour of building professionals. One result is that it brings up the question of responsibility vis-à-vis environmental risks, which has taken the form of a precautionary approach. To engage the bidder's responsibility, the order-giver is obliged to implement a reasoned approach that does not commit the future. To protect himself, the bidder must himself show that he has anticipated the consequences of his bid.

In a more concise way, all practitioners in the building sector will in the long term need to justify each of their choices, not just so as to guarantee a result, but to

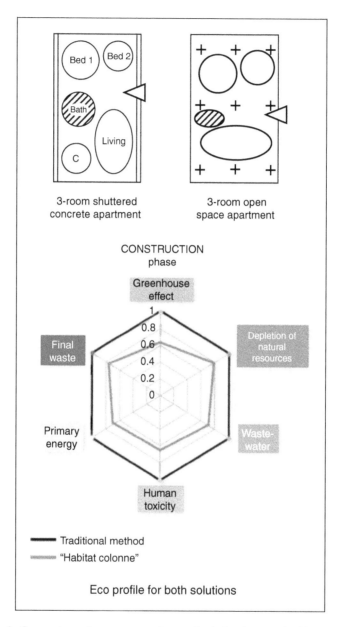

Figure 1 Comparison of two construction methods for the same building product.

show that the decision results from a critical analysis of the different possibilities. This therefore involves being able to explicate the reasons that led to the solution chosen. This obligation to provide information on the outcome of a practice is not so different from professionalism based on the capacity to explain a process rather than on the knowledge possessed.

Ecodesign is particularly well suited to this back to basics approach. It is by nature a comparative method that illustrates a comparison between two building solutions. Its object is not only to come up with a project's environmental profile, but to optimize it by visualizing the effect of introducing a building variant. This approach is more advantageous because it is carried out with "all other things being equal", since the comparison is made on the basis of the same functional unit (response to the same building programme).

The fundamental tool of ecodesign is its radar chart representation of the results calculated in the LCA. This illustration of the choices makes it possible to set up a dialogue between stakeholders. It is a vehicle for making joint decisions that fits in with a more participative approach to undertaking a building operation.

> *Ecodesign naturally leads to re-engineering and thus an innovation approach. Ecodesign means making informed choices, but it does not exempt the project manager from making a decision that only he is responsible for.*

The introduction of a comparative project analysis gradually results in a more critical approach to building methods. This is because examining projects more closely raises new questions about whether to continue repeating practices from simple routine habit.

For example, in France outside insulation now seems to be challenging the monopoly of inside insulation that was imposed by the structural principle of shuttered concrete. However, in the long term, the introduction of a carbon tax will also shake up certainties regarding load-bearing constructions. Cement manufacturers are already moving this way, and are working on "thermal concrete" and "green concrete" compositions.

The even longer term will involve a complete rethink of building methods, and ultimately a return to architectural models that minimize environmental impacts without increasing technical costs, including a review of the conditions for living in buildings. This formula means undertaking architectural research to organize living spaces to fit in better with this new construction economy.

This interweaving of technical, architectonic and economic parameters can only be resolved by looking for new paradigms. Together, these efforts will produce what is termed as "re-engineering", in other words, critically analyzing the action methods with the aim of organizing things differently and reaching a new level of financial optimization.

Ecodesign, in this perspective, could lead to a renewal of the industry combining an innovated product (the built architectural object) with a new conception of the productive organization (the building processes, i.e. the industry's technical procedures and organizational processes).

> *More than a methodology, ecodesign is a dynamic process. Ecodesign is part of a managed ongoing improvement.*

Ecodesign has been viewed in everything said above as carried out in a "stable state". In a given situation, it entails accomplishing a project while managing its environmental

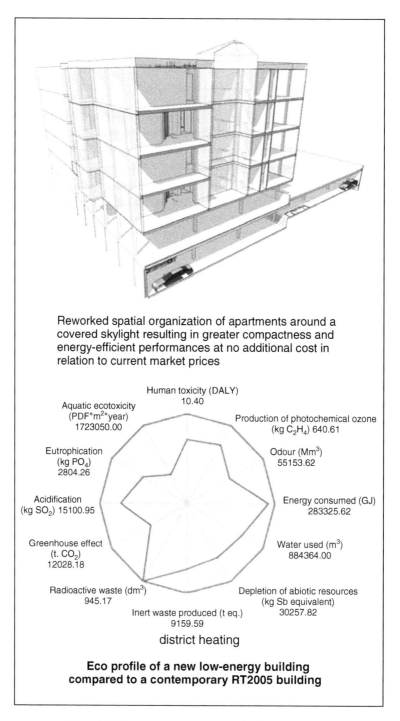

Reworked spatial organization of apartments around a covered skylight resulting in greater compactness and energy-efficient performances at no additional cost in relation to current market prices

Human toxicity (DALY)
10.40

Aquatic ecotoxicity
(PDF*m²*year)
1723050.00

Production of photochemical ozone
(kg C₂H₄) 640.61

Eutrophication
(kg PO₄)
2804.26

Odour (Mm³)
55153.62

Acidification
(kg SO₂) 15100.95

Energy consumed (GJ)
283325.62

Greenhouse effect
(t. CO₂)
12028.18

Water used (m³)
884364.00

Radioactive waste (dm³)
945.17

Depletion of abiotic resources
(kg Sb equivalent)
30257.82

Inert waste produced (t eq.)
9159.59

district heating

Eco profile of a new low-energy building compared to a contemporary RT2005 building

Figure 2 The capacity to create new building concepts.

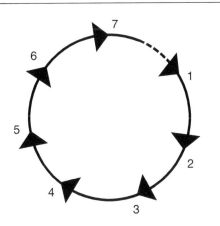

In the frame of a given operation programme, seven progressive principles should guide an ecodesign approach:

1- Launch an ecodesign approach
   voluntary commitment

2- Select resources with reduced impact
   environment and health (innocuity)

3- Reduce the quantity of resources used
   construction and operation

4- Financial mobilization of production
   study, design and construction

5- Manage impacts during use
   responsibility for use and maintenance

6- Optimize the product's initial lifespan
   upgrading and changed purpose

7- Optimize end of product
   deconstruction and recycling

It is the contractor's role to determine the level of principle that he wants his team to achieve.

Figure 3 The different set requirements of an ecodesign approach.

impacts. However, it can also be used to respond to set targets, in which case, it is worth taking a dual approach of establishing a performance and measuring the result.

In this way, ecodesign resembles an ongoing improvement. It can be practised while fixing an improved effort as shown in the diagram above.

The jurisdiction of this mechanism resides in the very nature of ecodesign, i.e. measurement. Improvements are only possible starting with measured observations. Although the input of these results corresponds to a simulation protocol, which is the LCA, the quantities it processes are tangible physical data, i.e. flows of matter and energy. This operation is also carried out per life phase, so that a detailed, progressive opinion can be established.

Construction is not difficult to measure because buildings are based on calculations. However, such measurements are justifications to conform to existing regulations. In ecodesign, the aim is not to verify whether the level of impacts is correct, but rather to attempt to minimize it. Ecodesign corresponds to voluntary commitment, not normalization.

In addition, an ecodesign approach based on the notion of a system avoids the simplification of a single-criterion, linear evaluation. It can thus only be part of a performance-based cycle, in other words: (1) specification of the level sought, (2) proposition of a satisfactory solution, (3) verification of the result in situ. What is important here is not to conform to a regulation, but to validate a way of operating that not only satisfies the order giver's expectations, but that provides a guarantee for future end users.

The status of ecodesign goes beyond that of static rules and corresponds to a momentum.

Ecotechnologies and building principles

Ecotechnologies are by definition structural solutions that correspond to ecodesign requirements. One question that immediately comes to mind is whether specific propositions exist, in other words, solutions specially adapted to ecodesign.

The aim of building using a limited volume of natural resources could result in what some experts term a "lightweight" construction. In fact, the issue is centred less on weight than on the overall balance sheet produced in the Life Cycle Analysis (LCA). In these conditions, the response is much more complex, since it requires discussing the entire building process, which is an updated version of the seventies debate between "closed industrialization" and "open industrialization"

The first term refers, at least in France, to a traditional, concrete-based building that rationalizes worksite tasks using appropriate equipment. The second refers to an on-site combination of manufactured components. The main difference lies in the place of manufacture and the supposition that tasks repeated in the workshop use fewer materials than those undertaken outside, which are subject to a range of hazards.

However, this kind of classification may no longer be relevant today. Although using lower volumes of concrete helps reduce greenhouse gases, it would be absurd to seek to totally substitute concrete for wood. The real issue here is mixing and combining technologies to improve the use of buildings. In ecodesign, no single solution exists, but rather a combination that fulfils specific criteria fixed by the contractor.

In these conditions, ecotechnologies are less intrinsically specific when they are the result of constructive reasoning that gives no more weight to any one material, with the sole concern of using the right quantity of the right technical procedure appropriate to each operation's specific context.

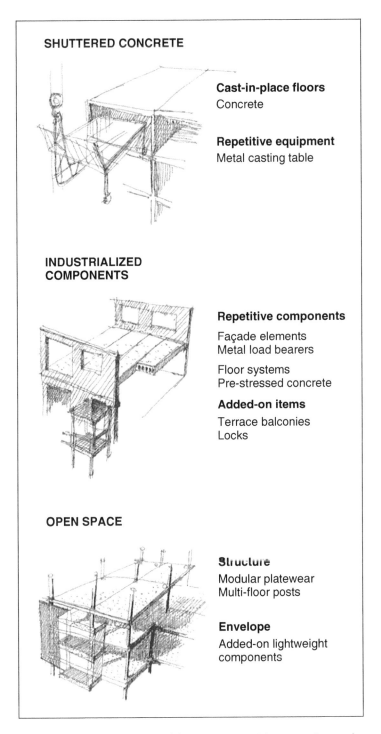

SHUTTERED CONCRETE

Cast-in-place floors
Concrete

Repetitive equipment
Metal casting table

INDUSTRIALIZED COMPONENTS

Repetitive components

Façade elements
Metal load bearers

Floor systems
Pre-stressed concrete

Added-on items

Terrace balconies
Locks

OPEN SPACE

Structure
Modular platewear
Multi-floor posts

Envelope
Added-on lightweight
components

Figure 4 Moving towards mixed, less mono-material construction work.

Ecotechnologies and the duration issue.

Ecotechnologies also bring up the question of their relationship to time: not so much their durability (absence of ruin) as their ability to match the financial time horizon fixed by the contractor. And from this point of view, two approaches are possible.

The first of these consists in adjusting the building's lifespan to the usage period planned only. In more figurative terms, the objective is to produce a "disposable" object. This kind of approach can be justified by at least two arguments: the first is to only invest the bare minimum, i.e. consider that a building does not need to survive beyond the time for which it was designed as long as its technical cost takes this commitment into account; the second is an observation of the performance level expected by end users, which continues to rise and thus rapidly makes all buildings obsolete. In these conditions, it is cheaper to construct a new building than to rehabilitate extensively.

Another reason is possible, and it is the one generally put forward by architects. A building also participates, beyond its users, to the collective domain as a part of a town. This urban heritage value could be justified by observing the various life cycles of each building component. The structure of a building has a longer life span than its coatings. In such conditions, it is possible to envisage a long-lasting structure housing different uses over time. The inside partition walls can be changed, renewed and modified as needed. Some architectural schools have even proposed urban structures to match a more short-term living environment.

In fact, both of these scenarios are possible and each may be justified based on the macro-economic choice criteria taken on by the order givers. It is however true that contributing to recreating an urban space is not currently valued as such, with the effect that the issue has not been debated to date. However, it will inevitably have to be dealt with in the near future.

How ecotechnologies are organized. Ecodesign corresponds to perimeters of different sizes.

The sector has gradually realized that it is not possible to consider a building for itself on an environmental level and that it should be understood within a broader perimeter. This is why the eco-neighbourhood concept has proved such a success. However, the need to measure impacts is still just as pressing and raises questions about the possibility of extending ecodesign at this dimension of a territory.

Neighbourhoods are initially defined as the sum of different buildings and an infrastructure grouping roads, networks and public facilities (e.g. lighting, street furniture). Each of these elements responds to a life cycle that can be modelled by an LCA. It is thus possible to imagine aggregating all of the corresponding impacts. As long as the data are available, it is possible to produce the environmental profile of a neighbourhood.

This process is currently underway: it has been prepared for European research projects and should rapidly result in a new supplementary module for the EQUER software programme, which is a tool specifically developed for the construction industry (buildings and infrastructure).

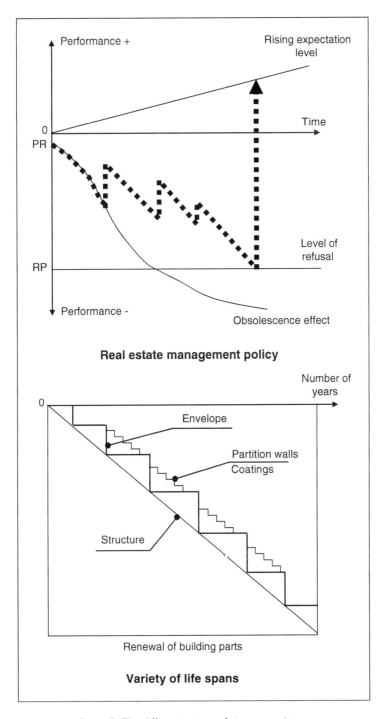

Figure 5 The different points of view on an issue.

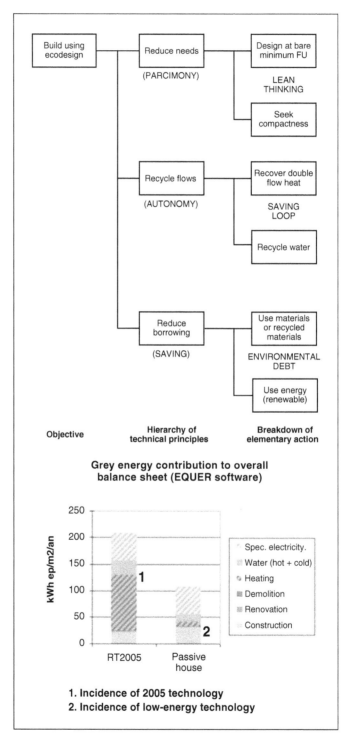

Figure 6   Relative weight of technology and logical structure of technical principles corresponding to ecodesign.

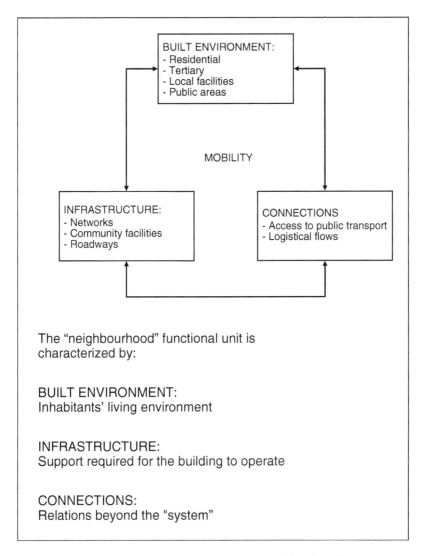

Figure 7   Model used to carry out LCAs at neighbourhood level.

However, creating a new neighbourhood from scratch is likely to be fairly limited when compared with rehabilitating certain existing neighbourhoods. This is the reason for working with industrial wastelands and upgrading certain suburban zones. The issue that arises with ecodesign is thus that of redeveloping the built environment. It is based on a two-fold approach that involves characterizing the initial situation and then the situation that will take its place. The difference between environmental profiles illustrates the efficiency of the planned project. The work principle is thus similar to that for new buildings but it uses two calculations.

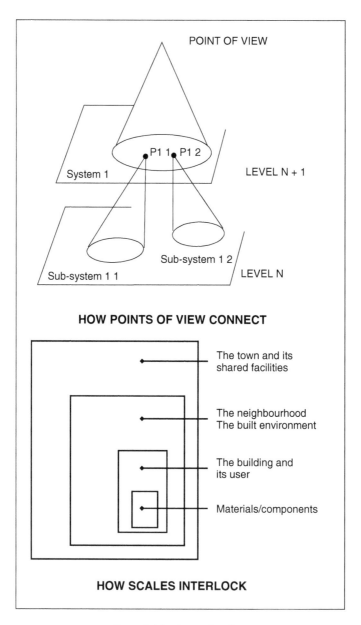

POINT OF VIEW

P1 1 P1 2

System 1

LEVEL N + 1

Sub-system 1 2

Sub-system 1 1

LEVEL N

HOW POINTS OF VIEW CONNECT

The town and its
shared facilities

The neighbourhood
The built environment

The building and
its user

Materials/components

HOW SCALES INTERLOCK

Figure 8 The issue of scales.

These two ways of extending a building's ecodesign illustrate the potential of this approach. This type of design nevertheless calls for rigorous procedures that are less time-consuming than they appear but yet highly dependent on a discipline that mainly involves collecting information.

Ecodesign applies on different scales. Modelling is currently being developed, involving correlating the extension of the physical perimeter with that of the functional perimeter.

Extending the ecodesign of buildings or road systems to a neighbourhood is therefore possible. However, it should go beyond a simple evaluation of static flows. The functional unit of a neighbourhood cannot be simply restricted to a geographical perimeter. New behaviour patterns also need to be taken into account.

This remark can be illustrated with an analysis of transit. In a set of buildings and road systems, it is justifiable to add up the flows resulting from the availability of these components. However, road systems are made up of a meshed network that supports traffic flows whose volume is constant at a point in time T (the need to move) but whose distribution (mode of transport) is unpredictable depending on users' decisions. To determine as closely as possible how this social relations unit works, it is therefore necessary to possess a probabilistic model on the use of certain types of transit.

In fact, each level of analysis is associated with the characterization of its functional unit, but the functions become more complex as the perimeter increases in terms of urban system. The functions do not simply add together by juxtaposition, they develop by connecting together in crossed processes. Two buildings involve transport flows towards the centre (hub) but if they are of a different nature (residential and commercial), they also generate an additional flow between them (catchment area).

A second development axis is opening up, which is the dynamic simulation of a neighbourhood. Currently, LCAs of installations can be used to envisage usage scenarios that will be duplicated throughout the entire operation phase. However, it would be more realistic to be able to develop these hypotheses in line with urban policy scenarios taking into account changes in distribution (changes in population, employment, cultural facilities, etc.) and also in occupation scenarios (ageing of users, modification in lifestyles, etc.). This research work is now under way and should be integrated into ecodesign tools in less than a decade.

Ecodesign is a component in the quest for efficiency.

Using an efficiency notion makes it easier to sum up the gains and limitations of ecodesign.

Put very concisely, all projects ideally seek to bring about the maximum number of usage values for a minimum number of suggestions. This is not so far removed from the saying, "do more with less". In its condensed form, it is perhaps the main objective of an approach that fits in with sustainable development. An object's efficiency is thus the relationship between the sum of the advantages obtained and the sum of the disadvantages generated over its entire life cycle. It is very simply illustrated by the graphic representation on the previous page.

In this diagram, ecodesign is seen as a working mode that involves managing impacts. Its aim is to reduce the burden resulting from making a building available – and at constant value. Ecodesign thus appears to be one of the components of efficiency. Some use the term eco-efficiency. This analysis has the merit of very clearly fixing the gain as well as the limitation of this approach. Ecodesign is only one of the parameters of efficiency.

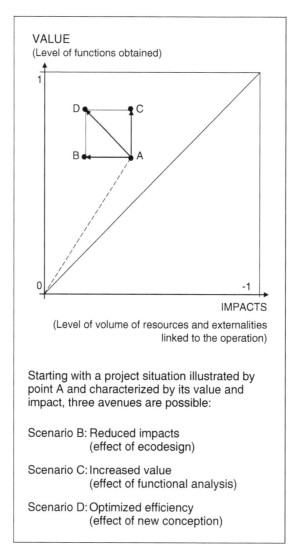

Figure 9 Different scenarios for conducting a project.

In fact, and this has been implicit throughout this publication, the art of construction is also based on a second vector, i.e. functional analysis, which makes it possible to work on the usage functions that the building provides a support for. This is why ecodesign has been closely linked to the notion of "functional unit", which is an inevitable prerequisite. Simply dealing with a project's environmental dimension is not therefore sufficient to fulfil the conditions of sustainable development.

The whole advantage of ecodesign is to be able to make a contribution to sustainable development without claiming to take its place.

Bibliography

Ecodesign is not as yet a totally autonomous disciplinary field. However, the subject has been tackled from different points of view and already given rise to numerous texts.

This selection centres on documents that together give a fairly comprehensive idea, without claiming to be exhaustive. This bibliography is organized into four groups:

Group A features documents that treat ecodesign in a general way applied to construction.

Group B includes publications that set out the main tools used in an ecodesign approach and in particular the standard-setting corpus.

Group C lists recent or current research that contributes to developing ecodesign in construction.

Group D lists the decision-making bodies that contribute to extending the use of ecodesign.

A. ECODESIGN AND CONSTRUCTION

BELLINI: "L'éco-conception état de l'art"

> A very concise overview of the concept. (Techniques de l'Ingénieur)

Bruno PEUPORTIER: "Eco-conception des bâtiments et des quartiers"

> The reference work on ecodesign in the building industry. Written by the EQUER software project manager. (Presses Mines ParisTech)

VINCI CONSTRUCTION: "Construire en éco-conception"

> Brochure outlining the main conditions for ecodesign in the building and public works industry (challenges, method and early examples).

EGF/BTP: "Recommandations pour l'éco-conception"

> Booklet summarizing the general principles of ecodesign and defining how it can be used by different practitioners working on a building operation.

EGF/BTP: "Recommandations pour des indicateurs d'éco-conception"

> This second booklet provides a more precise breakdown of the recommended indicators for measuring environmental approaches, both for projects and all construction company activities.

B. ECODESIGN TOOLS

A first sub-group covers the norms relating to the use of ecodesign. They all centre on LCA protocols in environmental management:

- ISO 14040: Principle and framework
- ISO 14044: Requirements and guidelines
- P 01.20.3: (AFNOR) Measuring a building's environmental performance

The second group sets out the conditions for establishing a functional unit:

- XP 50-151 (AFNOR): Expressing a functional analysis request
- Chr. GOBIN (Les Techniques de l'ingénieur): Functional analysis and construction

ACADEMIE DES TECHNOLOGIES:

> Leaflet with a very detailed look at life cycle analysis (LCA) issues.

C. DEVELOPMENTS AROUND ECODESIGN

L'ACV QUARTIER (ANR project):

> Creation of an LCA module on infrastructures to supplement the EQUER software programme on buildings (Mines ParisTech).

EFFICIENCE ET CONSTRUCTION

> A breakdown of the concept in the construction industry so as to respond the requirements of sustainable development.
> Chr. GOBIN – Techniques de l'Ingénieur

ECONOMIE FONCTIONNELLE ET CONSTRUCTION

> The move towards a new economic model for construction companies based on making a usage available.
> Chr. GOBIN – Techniques de l'Ingénieur

D. PARISTECH ECODESIGN CHAIR – VINCI

All of the points covered in this document will be progressively developed by the work of a ParisTech research chair associating laboratories from Mines ParisTech, Ecole des Ponts ParisTech and Agro ParisTech. It was launched in 2009 for a five-year period.

Around fifteen theses and post doctoral projects focusing on "ecodesigning buildings and infrastructures" will extend knowledge of existing tools and bring complementary information resulting from multi-disciplinary work to the attention of the scientific community.

.

Annexes

ANNEX 1. LCA APPROACH

Ecodesign has been tackled from the professional point of view of those involved in construction work. In this perspective, the central tool is clearly the life cycle analysis (LCA), an option that appears to be in operation.

The objective of this annex is to provide a more detailed presentation of the accepted principles of this method. This is because it involves a conventional representation that cannot in any case be validated by an experimental measure as is common in physical sciences, since the methodology's horizon is set at several decades.

Two components define a Life Cycle Analysis (LCA):

- The reference set of indicators chosen to inform choices,
- The compatibility protocol that converts flow quantities into impacts.

Terminology of environmental impacts

In the current state of the art, the following twelve indicators can be used to describe environmental impacts relating to buildings.

It is nevertheless worth pointing out that each of these indicators requires a calculation method whose perimeter and protocol are the subject of prolific debate within the scientific community given the recent nature of studies.

The selected indicators correspond to three levels of environmental target:

- Preservation of resources (energy, water, materials),
- Protection of ecosystems on global (climate, ozone), regional (forests and rivers) and local (waste and air quality) levels,
- Health related to the environment.

Explanation of the EQUER software chain

On a professional level, the objective is to obtain an aid for decision-making.

To achieve this, at the moment in France, only the EQUER software offers overall capacity covering all the components of a building, even though it has to make use of a European, rather than French, database (Ecoinvent).

Primary energy: The energy consumed over a building's life cycle is measured in terms of primary energy so as to take into account the different types of energy distributed (electricity, heat) on a homogeneous basis. The principle of a life cycle analysis consists in going back to the early phases of fuel extraction (crude oil, uranium, gas, coal, etc.) or other resources (e.g. hydroelectricity).

 Water used: This indicator is expressed in m^3 of water withdrawn.

 Abiotic resources: The resources of a river basin or geographic zone, still referred to as "in-place volumes", are the total quantities of matter present in discovered and to-be-discovered areas in the considered region, without any technical or financial consideration. Reserves are the quantities that have the potential to be extracted and exploited profitably in the near future. The transition from resource to reserve is characterized by the rate of recuperation.

 Final waste: Since the impacts of some waste treatment procedures that are not yet well known, an indicator has been defined whose value depends on the type of waste according to the cost of treatment.

 Radioactive waste: Waste is differentiated according to its activity and lifespan. Category A waste has low or moderate activity and is destined to be stored for around 300 years. Category B waste also has moderate or low activity, but contains very long-lasting elements, such as category C waste, which has very high activity. Type B and C waste must remain shut away for thousands of years. In the indicator that we use, we have added all of this waste together and expressed the quantities in dm^3.

 Greenhouse effect: This indicator is expressed in kg of CO_2 equivalent, and in our analysis we have considered a period of 100 years. The indicator represents the warming potential of various greenhouse gases converted into CO_2 equivalent.

 Acidification: This indicator corresponds to the acid rain phenomenon and the decline of forests.

 Eutrophication: This indicator is linked to the addition of substances acting as fertilizer in surface waters. These fertilizers cause an overproduction of algae, which deprives aquatic environments of oxygen when it decomposes, sometimes leading to the death of fish and other living organisms.

 Ecotoxicity: We have included several hydrocarbon values, but no value corresponding to pesticides, which do not directly concern the building sector. The indicators are expressed in m3 of water or kg of polluted soil.

 Human toxicity: The dose of pollutant received. One dose is the relationship between the mass of pollutant inhaled or ingested over a certain length of time and the weight of the individual.

 Formation of photochemical ozone: Some volatile organic compounds (VOCs) decompose under the action of the sun and contribute to the formation of ozone, which affects the respiratory tract. Depending on climate conditions, some towns restrict road traffic during ozone peaks to try and limit the "summer smog" phenomenon.

 Odours: Detection threshold of each odour, defined as the concentration when 50% of a representative sample detects the product.

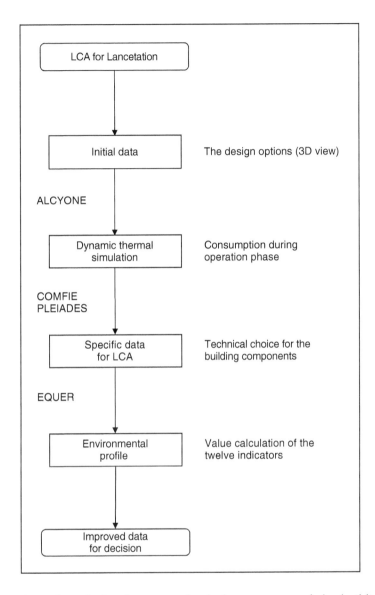

The notion of ecodesign has now clearly become part of the building sector. It involves taking into consideration environmental issues that are indispensible to putting together our living environment.

However, this method, which is of industrial origin, clearly shows that buildings are not the result of simply adding up technical rules. It is much more demanding. It engages all stakeholders in the industry and leads them towards using a new practice involving multi-criteria choices that are never unique.

The object of this publication is to review each of the stages in a building operation to illustrate the necessity of optimization and to observe the useful contribution that ecodesign and its tools can make.

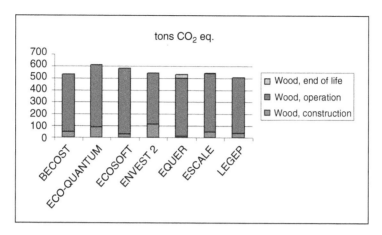

Figure 1   On European level, a benchmark has been put in place as part of the PRESCO network, reinforcing the use of EQUER.

It is therefore a handbook devised to start up a process of constant progress with the entire profession.

ANNEX 2. MEASURING AN ECODESIGN PROJECT MANAGEMENT

Many experts argue that an efficient project management is based on specific tools. To implement an eco-design management, we need several skills. Among them the Life Cycle Analysis Modelling is very useful.

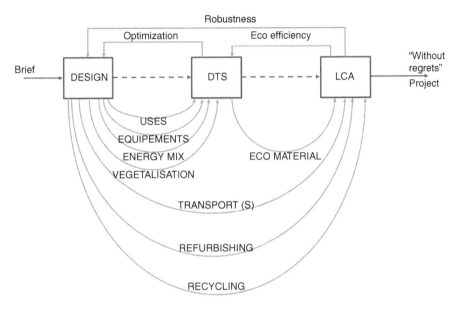

One figure can summarize all the capabilities supported by such a tool box.

The LCA software will be used to illustrate a recent project management: How to improve the construction choice for a new head office building.

This tool is accredited by BREEAM association.

The results have been established by the CES Research Center (MINES Paris Tech).

I The project

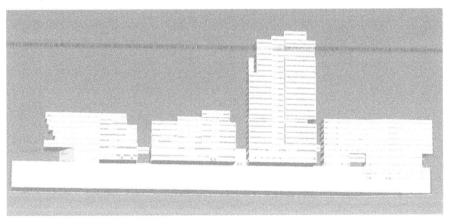

Figure 2 The project viewed from the north.

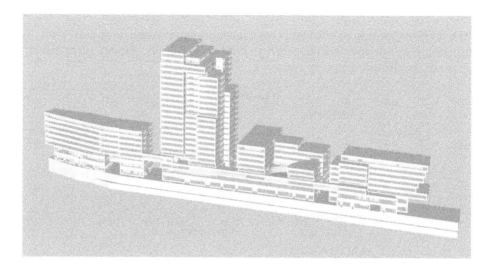

Figure 3 View from the south.

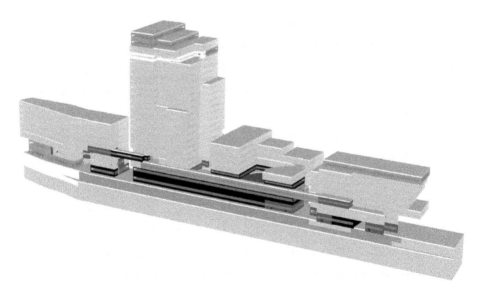

Figure 4 The several thermal zones.

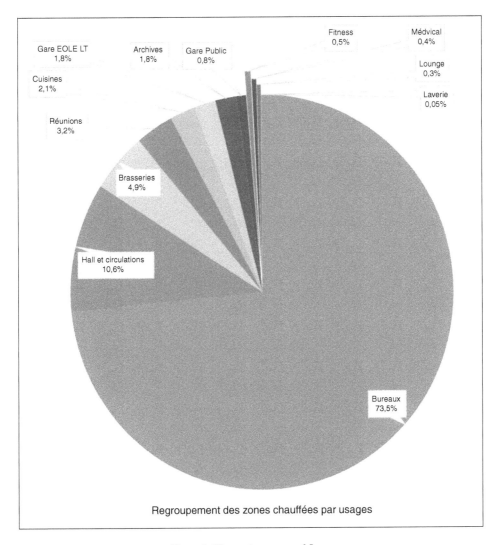

Regroupement des zones chauffées par usages

Figure 5 The various uses of floor.

This new building will be the Parisian head office of a large company (100 000 m²
and 60 000 m² conditioned). It is located above a new traffic exchange network. The
figure shows the different space allocation for the future activities.

The office heating and ventilation scenarios

Heure	Lun °C	Mar °C	Mer °C	Jeu °C	Ven °C	Sam °C	Dim °C
0-1h	16	16	16	16	16	16	16
1-2h	16	16	16	16	16	16	16
2-3h	16	16	16	16	16	16	16
3-4h	16	16	16	16	16	16	16
4-5h	16	16	16	16	16	16	16
5-6h	16	16	16	16	16	16	16
6-7h	16	16	16	16	16	16	16
7-8h	22	22	22	22	22	16	16
8-9h	22	22					
9-10h	22	22					
10-11h	22	22					
11-12h	22	22					
12-13h	22	22					
13-14h	22	22					
14-15h	22	22					
15-16h	22	22					
16-17h	22	22					
17-18h	22	22					
18-19h	22	22					
19-20h	22	22					
20-21h	16	16					
21-22h	16	16					
22-23h	16	16					
23-24h	16	16					

Hebdomadaire   % de ventilation   Ventilation DF Bureaux V05

Débit d'air nominal : 1.00 volume/heure

Heure	Lun %	Mar %	Mer %	Jeu %	Ven %	Sam %
0-1h	0	0	0	0	0	
1-2h	0	0	0	0	0	
2-3h	0	0	0	0	0	
3-4h	0	0	0	0	0	
4-5h	0	0	0	0	0	
5-6h	0	0	0	0	0	
6-7h	0	0	0	0	0	
7-8h	0	0	0	0	0	
8-9h	100	100	100	100	100	
9-10h	100	100	100	100	100	
10-11h	100	100	100	100	100	
11-12h	100	100	100	100	100	
12-13h	100	100	100	100	100	
13-14h	100	100	100	100	100	
14-15h	100	100	100	100	100	
15-16h	100	100	100	100	100	
16-17h	100	100	100	100	100	
17-18h	100	100	100	100	100	
18-19h	100	100	100	100	100	
19-20h	100	100	100	100	100	
20-21h	0	0	0	0	0	
21-22h	0	0	0	0	0	
22-23h	0	0	0	0	0	
23-24h	0	0	0	0	0	

To be able to model the building metabolism, first it is necessary to precise all the use scenario of equipment to deliver the performance wished by the end users described in the brief.

The second step is to characterize all the construction elements. It concerns the equipment and the built components. Two steps must be implemented: the geometrical and the physical characterisation.

Table 1 Envelope components.

Façades en béton

Composant	épaisseur (cm)	ρ (kg/m²)	λ (W/m · K)	R (m² · K/W)
Isolant générique	25	3	0.04	6.25
béton lourd	16	368	1.75	0.09
Total	41			6.34

Table 2 Partition components.

Cloisons en béton

Composant	épaisseur (cm)	ρ (kg/m²)	λ (W/m · K)	R (m² · K/W)
béton lourd	20	460	1.75	0.11

Table 3 Floors.

Plancher intermédiaire en béton

Composant	épaisseur (cm)	ρ (kg/m²)	λ (W/m · K)	R (m² · K/W)
béton lourd	20	460	1.75	0.11

Windows ratio

Pourcentage de baie vitrée par façade

Façade Nord	Façade Sud	Façade Ouest	Façade Est
30%	30%	40%	40%

Equipment

Ventilations utilisées

Type	Scénarios de ventilation concernés	Efficacité de l'échangeur	Consommation
Simple-flux (SF)	LT, Cuisine, Fitness & Parking	0	0.45 Wh/m³
Double-flux (DF)	Bureaux, Réunion & Public	0.8	0.45 Wh/m³

The geometrical options.

Concrete floor

Composante: Simple	Epaisseur (cm)	λ W/(m · K)	ρ kg/m³	CS Wh/(kg · K)	U W/(m² · K)	R (m² · K)/W
béton lourd	20.0	1.750	2300	0.256	8.75	0.11
Total					8.75	0.11

Basement floor

Composante: Simple	Epaisseur (cm)	λ W/(m · K)	ρ kg/m³	CS Wh/(kg · K)	U W/(m² · K)	R (m² · K)/W
Isolant générique	20.0	0.040	12	0.233	0.20	5.00
béton lourd	20.0	1.750	2300	0.256	8.75	0.11
Total					0.20	5.11

Vegetal roof

Complément	à compléter avec un Lamda très petit|
Valeur Up	− Up indicatif: 0.21 W/(m².K)

Composante: Simple	Epaisseur (cm)	λ W/(m.K)	ρ kg/m³	CS Wh/(kg · K)	U W/(m² · K)	R (m² · K)/W
Terres cuites (?n 1300 kg/m³)	20.0	0.460	1250	0.278	2.30	0.43
Plaques polyuréthanne expansé autre	20.0	0.050	40	0.389	0.25	4.00
béton plein (lourd)	20.0	2.000	2450	0.278	10.00	0.10
Total					0.22	4.53

Acoustic partition

Composante: Simple	Epaisseur (cm)	λ W/(m.K)	ρ kg/m³	CS Wh/(kg.K)	U W/(m².K)	R (m².K)/W
Plâtre + cellulose	1.3	0.300	1200	0.222	23.08	0.04
Laine de chanvre	6.0	0.039	25	0.390	0.65	1.54
Plâtre + cellulose	1.3	0.300	1200	0.222	23.08	0.04
Total					0.62	1.63

The physical characteristic

Lineic thermal bridges

All these data are completed by complementary features.

Nom	Classif.	Origine	ψ	$\psi 1$	$\psi 2$	$\psi 3$	
ITE 4.3.1 – Refend béton	4.3	CSTB	0.06	0.03	0.03	0.00	
ITE – Angle rentrant	4.2	CSTB	0.03	0.01	0.01	0.00	
ITE – Angle sortant	4.1	CSTB	0.16	0.08	0.08	0.00	
ITE – Pl haut / mur extérieur béton	3.1	CSTB	0.66	0.66	0.00	0.00	
ITE – Pl. bas sur VS / mur extérieur béton	1.2	CSTB	0.79	0.79	0.00	0.00	
ITE – Pl. intermédiaire béton	2.1	CSTB	0.09	0.04	0.04	0.00	

2   Optimization of building materials

Two aspects can be analysed. It concerns the insulation material and the concrete formulation.

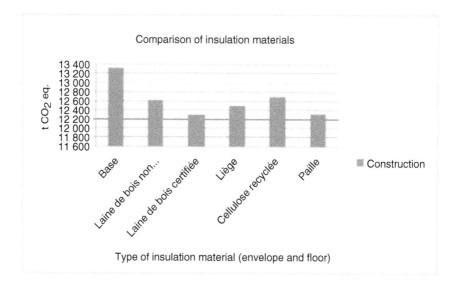

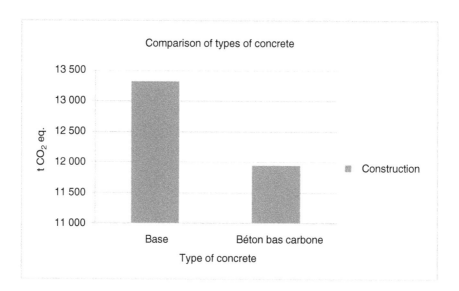

The results of the simulation can be visualized by the two next figures.

These data can be used by the project manager and help him to choose construction options.

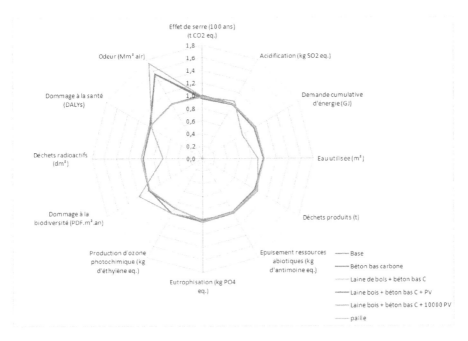

Figure 6 Lifecycle impacts of the various choices (50 years).

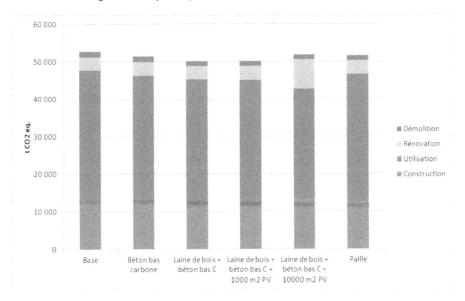

Figure 7 Detail of CO₂ balance.

3 Complementary optimization

As a building life cycle is long (80 years), it can be inspired to precise the impacts of each choice for each phase of the project duration.

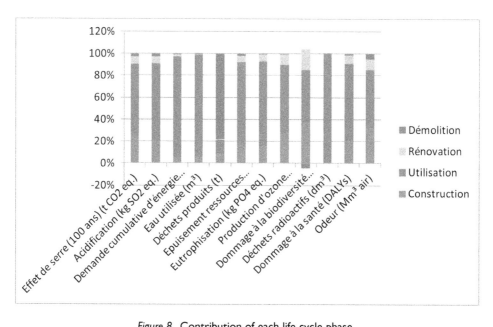

Figure 8 Contribution of each life cycle phase.

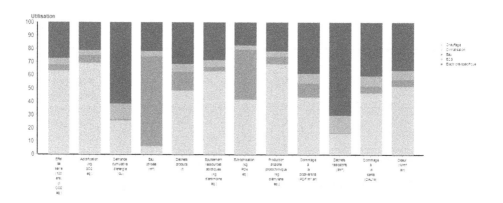

Figure 9 Sources of impacts.

In particular the detail of use phase is instructed. With such information, new options can be investigated.

3.1   A new energy mix

The figures are dedicated to the improvement by a renewable (mostly hydroelectric) consumption.

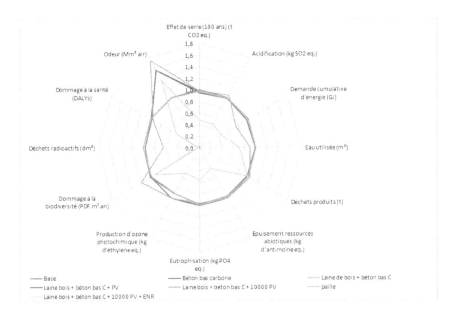

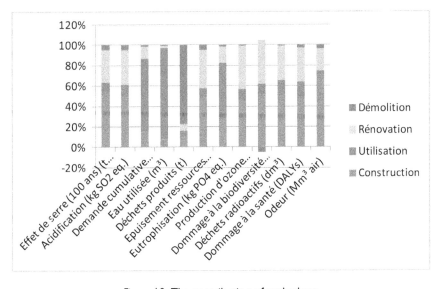

Figure 10 The contribution of each phase.

3.2 Other aspects for discussion

This first diagram shows the contribution of each building component.

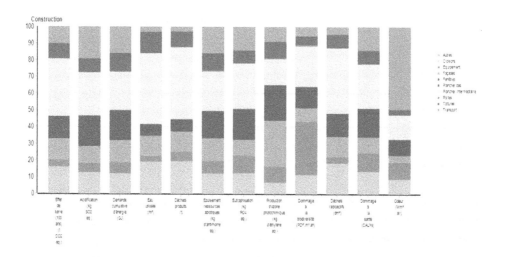

Figure 11 Impacts of construction phase.

Then it is possible to examine the improvement of specific process as photovoltaic panels.

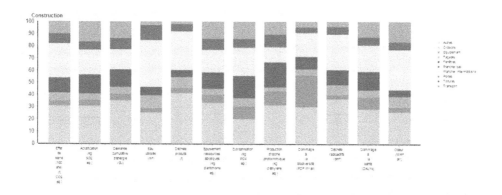

Figure 12 Impacts of the photovoltaic panels.

But all the life cycle stages must be investigated in particular the refurbishment.

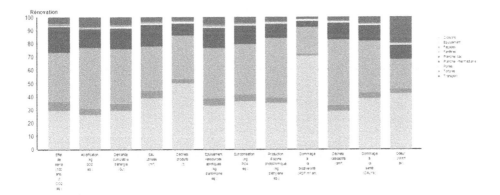

Figure 13 Impacts of grade up operation.

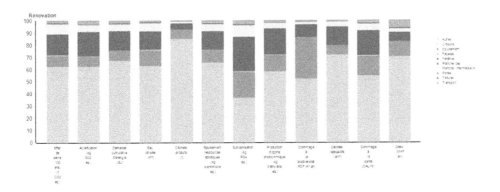

Figure 14 Impacts of grade up operation with photovoltaic panals.

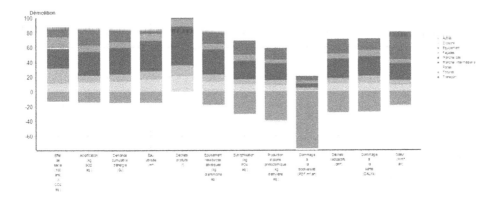

Figure 15 Impacts of deconstruction.

3.3   The final decision

The first figure illustrates the progressive optimisation of the project in terms of CO_2 emission.

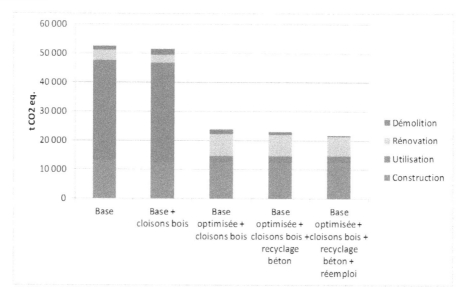

Figure 16 CO_2 balance of different possible solutions.

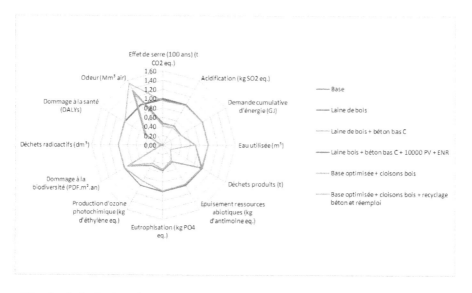

The final choice involves:

–   A wood fiber insulation
–   A low carbon emission concrets
–   Photovoltaic panels production and renewable energy use

Milton Keynes UK
Ingram Content Group UK Ltd.
UKHW051924141024
449569UK00027B/1345